COVERING DISASTER

COVERING DISASTER

Lessons from Media Coverage of Katrina and Rita

Ralph Izard and
Jay Perkins,
editors

Routledge
Taylor & Francis Group

LONDON AND NEW YORK

First published 2010 by Transaction Publishers
First paperback publication 2012

Published 2017 by Routledge
2 Park Square, Milton Park, Abingdon, Oxon OX14 4RN
711 Third Avenue, New York, NY 10017, USA

Routledge is an imprint of the Taylor & Francis Group, an informa business

Library of Congress Catalog Number: 2009042194

Library of Congress Cataloging-in-Publication Data

Covering disaster : lessons from media coverage of Katrina and Rita / Ralph Izard and Jay Perkins, editors.
    p. cm.
Includes bibliographical references and index.
ISBN 978-1-4128-1333-4
    1. Hurricane Katrina, 2005--Press coverage. 2. Hurricane Rita, 2005--Press coverage. 3. Journalism--Objectivity--United States. 4. Journalistic ethics--United States. 5. Journalism--Political aspects--United States. I. Izard, Ralph. II. Perkins, Jay.

HV636 2005 .G85 C68 2010
070.4'4936334--dc22

2009042194

ISBN 13: 978-1-4128-1333-4 (hbk)
ISBN 13: 978-1-4128-4582-3 (pbk)

Contents

Preface

The savage back-to-back attacks of hurricanes Katrina and Rita on the American Gulf Coast in the late months of 2005 produced a disaster of catastrophic proportions that challenged journalists, many of whom were numbered among the victims. The twin hurricanes devastated one of the nation's great cities, obliterated small towns, left thousands of people homeless for years, and left unbelievable physical and psychological damage among those who had called the area home.

This included journalists who had to do their jobs initially in virtual vacuums even as some of them worked to evacuate their families, dealt with losses of their own homes, worried about friends and family members, coped with the absence of most means of communication, and sought information from government officials who knew less than they did. Those journalists who did not live there faced problems, too, especially the human concern that grows from witnessing complete chaos, of seeing human suffering on a huge scale, and of dealing with rescue mechanisms that did not work or were nonexistent for days. They did their jobs under circumstances that made doing the jobs impossible.

This book is designed to analyze the journalistic responses to those challenges. What was the impact on journalism? On journalists? What did they learn, and what decisions did they make about the role of the media in their communities, and for the national media, the nation as a whole?

To seek some of the answers—to determine the "lessons learned," if you will—we present here the results of some of the work that emerged as scholars and professionals sought to gain understanding themselves. Each chapter, therefore, is a particularistic study of the human and professional coping mechanisms in the months and years following the storms. Some of the studies are quantitative, some are qualitative, and some are journalistic in their approach.

Furthermore, we have done some mixing and matching, combining the scholarly work with comments from professionals who covered the storms. Given the nature of both groups, it is not surprising that sometimes they agree and sometimes they disagree. But they all are thoughtful, and each provides a piece of the puzzle.

The combined impact of these individual discussions, we hope, carries this book beyond what appears to be a focus on crisis coverage. In spite of many errors, we believe the journalists along the Gulf Coast were forced to rediscover and emphasize those journalistic purposes and techniques that have long been the hallmarks of greatness. In that sense, we seek to provide some depth to a discussion of the qualities of good journalism. We hope the weaknesses of the coverage will provide lessons that thoughtful journalists will use. We hope the strengths of coverage will be remembered and carried into the daily activities of journalists of the future.

That longer-range question of whether the answers found in the midst of the storms will serve as catalysts for better journalism remains unanswered, and the trends are not always encouraging. Let us hope the lessons will not be forgotten now that the winds have died down, the buildings are being renovated and rebuilt, and the names Katrina and Rita are receding into historical memory.

We are most grateful to those authors who allowed us to rework their scholarly efforts, to combine their work with discussions from the field. We wanted to use the studies produced by these scholars and the responses of their journalistic colleagues in an analysis that allowed that research and those professional experiences to work together. It was a construction site that used all available resources.

It's not over yet, whenever you read this book. Both physical and mental repairs continue to be made, and the process of mourning for those lost may never end. Scholars continue their studies, and some journalists continue to seek understanding. It is our hope that this book contributes to all of these efforts.

As is the case with any project of this size, it may not be possible to recognize all of those who contributed to whatever success this book may achieve. Those people are numerous, ranging from those who physically contributed to these chapters to those who provided continuous advice, counsel, and information.

In particular, though, we share our thanks and admiration for the contributions of Erin Coyle, Melody Sands, Jennifer Kowalewski, Barbara Raab, Masudul Biswas, Amy Martin, and Kirsten Mogensen. It would not be the same without them.

—RIZ and JLP

1

In the Wake of Disaster: Lessons Learned

Jay Perkins and Ralph Izard

> *We not only have the responsibly to inform viewers about the facts of a situation during a tragedy, but we also have a responsibility to understand how the information we are presenting is going to emotionally impact our viewers and let this guide how we actually present this information.*
> —*John M. Gumm, WKRC-TV in Cincinnati.*

It's not cynical to note that journalism thrives on bad news. One definition of news, after all, is that which is not normal. Clearly, the onslaught of hurricanes Katrina and Rita was not normal. And, just as clearly, both print and television thrived from a story painted on a canvas that stretched from Alabama to Texas, with devastating scars gouged into Mississippi and Louisiana as far inland as 80 to 100 miles.

The twin anomalies of nature provided the print media with more than they could easily handle. And they gave television a moment in which to not only thrive but to prosper. For nothing can capture devastation better than pictures. Nothing can capture human suffering better than images, whether moving or still. And nothing can capture moral outrage better than a visual depiction.

It had only been four years since journalists had to deal with a story as massive as this. Clearly, the lessons learned from dealing with the terrorist attacks in New York, Washington, and an isolated field in Pennsylvania were still fresh in the minds of reporters. But this story was different. Not only was the damage caused by Katrina and Rita far more extensive, the story was purely domestic—and predictably expected. There was no foreign enemy to rally against,

1

no shadowy figures plotting the destruction of the United States from abroad. This was a natural disaster, an act of nature, a storm on a path of destruction that had been charted for days. Three states were badly damaged, dozens of beachfront and low-lying communities damaged or laid waste, thousands of families displaced. And that was before the levees in New Orleans gave way.

The only surprise was that the levees—those manmade piles of dirt, rock, concrete, and steel designed to keep low-lying New Orleans above water, failed miserably. And the huge pumps that supposedly would pump out the rainwater, and perhaps even some seawater blown in by the hurricane, did not work.

The human suffering that resulted from Katrina and Rita went a long way toward reshaping domestic politics. No American city had ever been so badly damaged as New Orleans. Earthquakes that have hit San Francisco and Los Angeles since the age of television pale in comparison. The attack of 9/11 on New York and Washington were measured in square blocks—not in square miles.

"The 'blast zone' of Hurricane Katrina was bigger than that of Hiroshima and Nagasaki," noted *Wall Street Journal* writer and editor Ken Wells. "Historically, not since the plagues of London (1665-66) has a great city been so depopulated. Reporting was done under very extreme circumstances."[1]

Not only was the scope of damage huge; the logistics were nearly impossible. The storm closed major airports in the region, wrapped trees around broken power lines, turned interstate highways into obstacle courses, and uprooted well over a million people. The city of New Orleans was turned into a lake that was 20 feet deep in some places. Boats became the only way to get around—and boats were not on anyone's planning list when the event began.

The strain of the journalistic effort was intensified for local journalists by the fact that they were victims themselves. They were worried about more than their professional responsibilities. They were frantic about their families, their neighbors, and about their homes, many of which were destroyed by the winds and waters. Many lost contact with their loved ones. It was not that they did not know whether their families were safe—they did not know in some cases where they were.

And, journalistically, all the correspondents who covered Katrina were on their own. There were no newsrooms to provide comfort and support a few miles up the street, no rooms for sleep, no convenient power outlets to recharge batteries. Many of the reporters covering the flooding of New Orleans were working out of Baton Rouge, some 80 miles northwest. Others could find no place closer than Lafayette—another 40 miles up Interstate 10 from Baton Rouge. And a great number of them were bunking on couches and sleeping on floors even in those towns.

And yet they did the job. Most would say that, overall, they did the job well in spite of numerous mistakes. Journalists provided information when information was almost impossible to get. They informed a beleaguered public who needed facts that literally could save their lives. They mourned the loss of life and the suffering of those who faced unbelievably harsh and trying personal circumstances. They harangued government officials whose responses to the physical suffering were universally considered to be inept.

They did not stop at covering the unfolding events. Both print and broadcast media did not pack up their notepads or their cameras and head off into the sunset when the adrenalin surge had ended. Instead, they showed the promise that their founders—those who shaped the early days of media in this country—had hoped for.

Then the journalists who covered hurricanes Katrina and Rita continued their analysis of what happened, the impact to property and to human lives, for months and years after the events. Print and broadcast reporters alike looked at broader issues, especially how government performed and why it did not perform as well as was expected. They showed they could continue to report, investigate, and analyze the longer-term situation and even reshape the political landscape.

Reporting from the Heart

The difference in coverage of the Gulf Coast hurricanes occurred because journalists abandoned their traditional impassive approach to the news and allowed their feelings to show in their reporting. As Geneva Overholser, director of the School of Journalism at the University of Southern California's Annenberg School for Communication, noted, for example, "Reporters (most notably Anderson Cooper of CNN) showed their hearts."[2]

There was emotion aplenty when Katrina and Rita attacked America's Gulf Coast. The pictures of families trapped on their roofs under the hot sun, surrounded in all directions by water 10 and 12 feet deep, was emotional from the start.

It was such emotion that prompted Ted Jackson, senior photographer for the *Times-Picayune* in New Orleans, to decide he could not take a picture of any person he could not help.

NBC News anchor Brian Williams showed the personal impact of what he witnessed when he declared on air: "I think of the faces, I think of the babies, I think of the elderly, I think of the people who, just a few days earlier, had dignity, had their lives, weren't defecating where they stood, weren't reusing their children's diapers.... I felt I had a privilege, an honor, of representing them. It was an honor to be with them in the Superdome. It was an honor to represent their interests, to do their pleading on national television."[3]

And it was feeling beyond dispassionate journalistic logic that led John M. Gumm, then a meteorologist for WWL-TV in New Orleans (now with WKRC-TV in Cincinnati), to determine the facts alone were inadequate in reporting on the approach of Hurricane Katrina. He knew that prior to Katrina, the people of New Orleans had evacuated in the face of several hurricanes that eventually turned away from the city.

"So how do I convince my viewers Katrina is not going to turn like the others?" he asked.

> How do I give the worst possible news to viewers in a responsible way? In my mind, I knew they would not want to believe me. With my wife nine months pregnant, even I did not want to believe this hurricane was coming. But it was. How would I convince my viewers?
>
> Well, instead of over-focusing on meteorological information, I decided to talk to my viewers like I would talk to friends. I told them I did not want to believe it was coming either, but it was, and there was nothing we could do about it. But what we could do was prepare for it and evacuate to protect ourselves and our families. We were not helpless.
>
> Covering the approach of Katrina was certainly the most important work I have ever done in my life because I was not only presenting critical information to my viewers, but I was also able to reach them on a personal level, and that made all the difference in the world in their ability to understand and use the information I was presenting.

Psychiatrist Frank Ochberg, a founder of the Dart Center for Journalism and Trauma, said it is healthy for journalists to show emotion.

"A journalist who goes to war has as much emotional casualty as the soldier who goes to war," Dr. Ochberg said. "The culture of journalism has been to ignore this, to deny this, to treat it with alcohol and bravado and a certain amount of contempt for the journalist who admits a problem."

And while he noted that some argue journalists must be cool under fire, "cool is one thing and cold is something else."[4]

When Emotion Turns from Sorrow to Outrage

The shock of seeing an American city destroyed and thousands of human beings suffering and dying created initial emotion. But the shock of seeing nothing being done about it quickly turned that emotion into anger. Citizens and reporters alike saw firsthand how inefficent and inept government really was. It was not just a failure of one agency or one branch of government. It was a catastrophic failure from top to bottom, from the sheriff on the street to the bureaucrat in Washington.

And while America's Nero fiddled, people were dying, trapped in their attics, buried in the rubble of buildings, and bedridden in hospitals with no power and no ability to help patients.

"You got a kind of journalism of witness or journalism of outrage," said Bruce Shapiro, executive director of The Dart Center for Journalism and Trauma, in discussing coverage of hurricanes Katrina and Rita.[5]

Many journalists clearly abandoned the concept of fair-and-balanced coverage and became advocacy reporters, telling the people in no uncertain terms what they were witnessing. They were no longer reciting information. Newspaper and television reporters began providing answers, naming names, and pointing out where the problems were. Nowhere was this more evident than when Anderson Cooper of CNN interrupted Sen. Mary Landrieu, D-La., who was congratulating the Senate leadership for calling an "unprecedented" special session to extend funding for the Federal Emergency Management Administration (FEMA) and the Red Cross.

"Excuse me, Senator," Cooper exclaimed.

I'm sorry for interrupting. I haven't heard that because for the last four days, I've been seeing dead bodies in the streets here in Mississippi, and to listen to politicians thanking each other and complimenting each other, you know, I've got to tell you, there are a lot of people here who are very upset and very angry and very frustrated. And when they hear politicians…thanking one another, it just—you know, it kind of cuts them the wrong way right now. Because literally, there was a body on the streets of this town yesterday being eaten by rats because this woman had been laying in the street for 48 hours. And there's not enough facilities to take her up. Do you get the anger that is out here?[6]

BBC correspondent Gavin Hewitt said such outrage is an emotion that stands out in television coverage because it is rare.

"Most reporters shy away from letting their emotions show," he said. "We all felt a sense of outrage that this should not be happening. Others felt that same emotion when they saw bodies that lay uncollected day after day. Outrage should be used sparingly and should never slide into anger. Outrage is at its most effective when it is based on compassion; the sense that one is speaking out on behalf of ordinary people."[7]

Seattle Times TV Critic Kay McFadden said the outrage expressed in coverage of Katrina showed America's *passion for passion*.

"What we liked, overwhelmingly, was feeling. Emotion spiked across all formats, from the weeperoo reality series *Extreme Makeover: Home* to psyche-stripping dramas like *Grey's Anatomy* and *House* to that memorable August week when Katrina forced reporters to abandon objectivity and voice outrage," she said.[8]

The University of Southern California's Overholser said the outrage went over well with the public.

"Readers and viewers loved it," she wrote. "Even months later, the *New York Times* on April 10, 2006, quoted a reader in New Orleans, saying: 'These writers are energized and passionate.' She wasn't a big fan of the paper before Katrina, she said, but now if she misses a day, 'I feel so out of touch.' A headline accompanying the story summed it up: 'Coverage driven by shared grief over losses and hope for rebuilding.'"[9]

Has Objectivity Outlived Its Usefulness?

Such journalism clearly has implications for the traditional journalistic practice of objectivity. Dispassionate reporting, fairness, and balance have been defined as essential ingredients of the journalistic creed since the rise of mass media created larger audiences composed of more diverse groups. Journalists have been taught they must be fair and balanced—and, at the same time, they must seek the truth. Yet the truth of hurricanes Katrina and Rita was not fair or balanced and certainly not dispassionate. Truth never is.

Tom Rosenstiel, director of the Project for Excellence in Journalism, noted a parallel between coverage of the hurricanes and the tenor taken by journalists in covering the civil rights movement, the Vietnam War and Watergate in the mid-twentieth century.

"I don't think one hurricane has the same galvanizing effect of a generation of incidents," he said. "But (after Katrina) you had street reporting suggesting that the official, whitewashed versions of some things were untrue. And that changed the tone of coverage."[10]

Jim Amoss, editor of the *Times-Picayune* in New Orleans, said this need to tell it like it is clearly influenced his newspaper's Pulitzer Prize-winning coverage of the flooding of New Orleans.

"It's very difficult to get at a truth as complex as what happened to this American city and its destruction," he said. "So I don't like 'on the one hand, on the other hand' journalism. We have to call it as we see it and not tiptoe around the subject."

The *Times-Picayune* clearly abandoned any pretense of objectivity in its coverage of Katrina. After all, its staff members were victims of the storm themselves. That made them true experts, and it was natural that their newspaper would become a strong advocate for New Orleans and a strong critic of the federal and state responses to the plight of the people. And even when that disaster was two years in the past, the newspaper had not changed its attitude that advocacy is more important than objectivity, that truth is the ultimate goal.

"I think we've become much less of a tiptoeing-around-the-subject newspaper, that we get to the point, that we call things as we see them," Amoss said. "We don't hesitate to zero in on problems and to criticize leaders when they fall short."

That many journalists abandoned traditional objectivity in covering Katrina and Rita is well accepted. It wasn't just Anderson Cooper getting mad at Sen. Mary Landrieu, and it wasn't just Shepard Smith on rival FOX News showing his ire. As Katrina wore on, the lack of objectivity grew along with the frustration. Overholser noted others among the many examples.

"Chris Wallace of FOX News asked Homeland Security's Michael Chertoff: 'How is it possible that you could not have known on late Thursday, for instance, that there were thousands of people in the convention center, who didn't have food, who didn't have water, who didn't have security, when that was being reported on national television?'" Alessandra Stanley of the *New York Times* wrote a piece headlined "Reporters Turn from Deference to Outrage."[11]

Ong Soh Chin, writing in the *Straits Times*, said American journalists "abandoned the profession's golden rule of objectivity.... CNN's Jeanne Meserve wept openly on TV while her colleague Anderson Cooper, as well as Mr. (Tim) Russert (NBC News) and ABC's Ted Koppel, grilled inept officials and demanded answers." It suddenly seemed like journalists were rolling up their sleeves and deliberately influencing the news even as they were reporting it.[12]

But Cooper, who garnered much of the credit—and most of the flak—for showing emotion on screen, warned about the dangers in carrying this type of thinking too far. He said his initial outburst "wasn't a conscious choice" and he certainly was not abandoning his belief in a form of objectivity.

> I believe very much in being objective. I don't believe in wearing my politics on my sleeve. I don't take sides. I know it's a popular thing in cable news these days to take sides. I just don't do it; nor will I ever do it. Viewers are smart enough to make up their own minds. They don't need, you know, an overpaid, blow-dried anchor like me to be telling them what to think or how to think.
>
> But I do think that there are cases when the least—the least our representatives can do is answer questions. Not give responses to questions, actually give answers to questions. And I think when you're in a situation where you're being told one thing and yet you're seeing all around you the other... I think clearly there's nothing wrong with confronting people with the facts that you are seeing. [13]

Likewise, as he witnessed the outrage boiling over, Ong Soh Chin, senior writer of the *Straits Times*, issued a similar warning. "2005 was the year the stakes went up in journalism," he said. "Trying not to get burned on them is the new challenge."[14]

Regaining Media Relevance

But these trends came at a time when journalism clearly needed to consider its role in society and to analyze its practices. Precedence is available. The emotion-packed journalism on television in the aftermath of the hurricanes clearly resembled the "righteous advocacy journalism" made famous by Edward R. Murrow during his career with CBS News in the early days as broadcasting developed its news and public affairs programming.

That sense of advocacy was lost by television sometime in the 1970s, and the years following 9/11 produced among all media a much-criticized brand of journalism that failed to even ask the tough questions. The concept of the journalist as one who dared "speak truth to power"[15] gave way to the concept of the journalist as entertainer. News values gave way to personality-driven coverage. Reporters began, in the words of the late author Michael Crichton, "selling the sizzle without the steak."[16] Reporting in both print and broadcast became bland, questions less confrontational, the audience for news smaller.

This, in part, could have resulted from the increasing complexity of society. Reporters often were generalists, jacks-of-all-trades and masters of none, yet they now face issues that demand the understanding of a specialist. Faced with little true knowledge of the issues, they resort to general questions and accept general answers. Confronted by complexity, they became ever more reliant on the version of events woven by the spin-doctors of political communication and public relations.

And they end up asking the type of questions Crichton complained about a decade earlier that dictate an answer or "assume a simplified, either/or version of reality to which no one really subscribes." The result was that large segments of the American population thought the media "were attentive to trivia and indifferent to what really matters," he said, adding that many in the public believed the media do not report the country's problems, but instead are part of them.[17]

This contributes to what many believe is an increasing irrelevance of the media to the public. That trend is especially important when coupled with the rise of new forms of self-generated media, such as Twitter, blogs, and YouTube, which have emerged as competitors for the shrinking free time resultant of modern life.

The decline of the news media's influence has been documented on numerous occasions in numerous ways. Some have argued, as did Crichton, that the media brought it on themselves. Others argued that changing lifestyles and economic circumstances made media less important to the audience. But no one has argued that it has not occurred.

The thirty-year precipitous decline in readership of newspapers and in viewership of television could be seen in past years as newspapers moved into joint operating agreements and then into single-ownership markets. The move this past year has been toward online distribution and away from home delivery of a paper product.

Television has not been any more successful. Live viewership dropped 10 percent in 2007, Reuters reported. Even when time-shifting with digital video recorders was factored in, the big four networks lost 5 percent of their viewership. Local television stations could be facing an even more dismal future, the *Wall Street Journal* has noted, if the networks try to stay alive by funneling their shows directly to cable instead of sending them to affiliates.[18]

The media's response to Katrina and Rita gave journalists a brief reprieve from declining audience and decreasing importance. People tuned in—and they stayed tuned in—in great numbers. The hurricanes showed that the media are not dead and hardly close to extinction. They demonstrated that journalism still has an important role in society. But it is clear that the media should learn the lessons of their coverage of Katrina and Rita. They need to remember those practices that created a strong bond between themselves and their audiences.

Accuracy, Accuracy, Accuracy

One lesson clearly shown by Katrina and Rita coverage is that the media cannot abandon their mantra of "accuracy, accuracy, accuracy." That mantra becomes all-consuming in the midst of chaos. The major media function in a crisis is to provide clear, precise, accurate, and timely information to thousands of people who must make decisions about their own safety. Often, there can be no margin of error.

Even though media overall did well in covering Katrina and Rita, numerous examples indicate the coverage at times was incomplete or off-target, sometimes factually inaccurate, often overly sensational and perhaps even insensitive. Two years after the hurricane, columnist Jonah Goldberg took an extreme position when he criticized the continuing celebration of journalism's good work by asserting that "Katrina represented an unmitigated media disaster as well."

"Few of us can forget the reports from two years ago," he said.

"CNN warned that there were 'bands of rapists, going block to block.' Snipers were reportedly shooting at medical personnel. Bodies at the Superdome, we were told, were stacked like cordwood. The *Washington Post* proclaimed in a banner headline that New Orleans was 'A City of Despair and Lawlessness,' and insisted in an editorial that 'looters and carjackers, some of them armed, have run rampant.' FOX News anchor John Gibson said there were 'all kinds of reports of looting, fires and violence. Thugs shooting at rescue crews.'"[19]

Goldberg asserted that these reports actually hindered rescue efforts, a point accepted by Robert Mann, then on the staff of Louisiana Governor Kathleen Babineaux Blanco. The false stories—especially those that suggested that rescuers were the targets of violence—almost certainly hindered the rescue efforts by discouraging some first responders, relief workers, bus drivers, and others from offering assistance in the crucial days after the storm, Mann said.

It may be argued that the catastrophic nature of the hurricanes and the chaos in the aftermath made errors of fact inevitable. That's true, but it's not the individual cases of inaccuracy; it's the question of why they occurred. No self-respecting reporter or editor could condone some of the answers to this question—dependence on rumor without adequate checking, too much haste, inadequate caution, an overwhelming attitude of competition that makes the filing of a story now more important than checking it out, automatic acceptance of information—especially criticism—gained from a public or political official.

Mann agreed that his own state government shared the blame for problems of coverage, but he added: "Too often, the attitude of these reporters was cavalier, as if accuracy in television news was merely a consequence of evolution or trial and error. Live television and the immediacy of the crisis, it appeared, caused some reporters to repeat rumors and hearsay as fact."

Perhaps the most public, large-scale, journalistic error occurred when many in the national media noted that Hurricane Katrina had passed and New Orleans had "dodged the bullet." Hindsight tells us this was not the case. The worst came later with the flooding of 80 percent of the city.

Local journalists, print and broadcast, knew this was a possibility, perhaps even likely, and they were pre-positioned around the city and throughout the region in advance of the hurricane. The visitors, on the other hand, tended to choose the most obvious locations from which to operate. They cannot be faulted for seeking safety and ready access to official sources. In the case of Hurricane Katrina, this was the relatively higher ground of the French Quarter. However convenient, it turned out to be the wrong site to get the real story. The local journalists were on the shores of Lake Pontchartrain, at the 17th Street Canal, the London Avenue Canal, and the Industrial Canal when the levees were breached and the water poured into the city.

It's true that journalists for national organizations focus on somewhat differing functions in a crisis, to provide a broader perspective for a broader audience. They covered the immediate story of human suffering in New Orleans, for example, that, in fact, did occur at the Superdome and the New Orleans Convention Center. But the networks do provide information, and on the cause of the long-term destruction, they played catch-up because they did not understand the geography of the region.

Plan But Be Inventive

A second lesson is that the experiences of covering the Gulf Coast hurricanes reinforced what good journalists know already:

The job of a journalist cannot end with witnessing and describing. It must become an intellectual exercise and, like all intellectual exercises, it must begin with the ability to plan and end with the ability to discard all plans if necessary.

Coverage begins before the crisis with journalistic understanding of what is possible—and it must begin in ways that garner public interest and attention. Journalists must issue warnings, as the *Times-Picayune* in New Orleans did in June 2002 in a five-part series about the likely disastrous impact of a major hurricane hitting New Orleans.

Yet, at some point, planning and prediction must give way to inventiveness and intelligence. The *Times-Picayune* and New Orleans' WWL-TV did not plan to operate newsrooms from the Manship School of Mass Communication at Louisiana State University in Baton Rouge more than 80 miles away. The *Biloxi Sun Herald*, which shared the Pulitzer Prize for Public Service Reporting with the *Times-Picayune* in 2006, did not expect to be publishing its newspaper in Columbus, GA, and trucking it in daily. And few national reporters anticipated needing a boat, a four-wheel-drive vehicle, or satellite phones to cover disaster relief efforts in the middle of an American city.

But reporters are inventive and well trained. So it was no surprise that Chris Adams, at the time a reporter for Knight Ridder's Washington bureau, intentionally violated his company's thrifty rental policy to grab the last Hummer available as he headed toward New Orleans. It was no surprise that he and the reporters he worked with had three cell phones connecting to different services plus one satellite phone. It was no surprise that when none of the phones would work, he and others found other ways to transmit their prose and their pictures.[20]

Interestingly, the hurricanes forced some journalists to understand that the rise of the Internet was not a competing force but a complementary force. Graphics and interactive databases were used to supplement traditional coverage. Items that could not fit into a television format found their way to television websites. Newspapers with crippled facilities turned to web presentations and later expanded their day-to-day routines to include the Internet.

Only Relevance Can Break through the Clutter

Katrina also taught that bigger and better stories may well be the key to success in journalism. The media clutter that, sadly, is American journalism (and American society) today continues to increase at a rapid rate. The traditional media today compete with thousands of other voices. The result, as social anthropologist Thomas DeZengotita, wrote shortly after 9/11, is a "clogged, anesthetized, numb" society.[21]

Every moment of our lives is now filled with media offerings, whether newspapers, radio, television, iPods, MySpace, Facebook, or YouTube. We have at our fingertips the ability to read hundreds, if not thousands, of mainstream newspapers from around the globe. We have at our fingertips the ability to watch recorded television newscasts from other countries on our television or on our iphone, to bundle our favorite television show for watching while walking for exercise, to timeshift the media we do not have time to watch into another spot in our busy schedules. We share our innermost thoughts—and our common complaints—with friends and total strangers through the instant messaging service provided by Twitter.

In the midst of all these messages, the traditional media will compete only if they reach directly into the minds and hearts of those publics for whom they exist. Citizens knew they needed information during and in the aftermath of hurricanes Katrina and Rita. And, happily, journalists responded in such a way that they gave that information and demonstrated that they could once again be critical to society. The media took on the roles of defenders of the public and prosecutors of government inefficiency, and they found ways to become leaders to those who read them, view them, listen to them.

Most Important May Be the Tone

Many cities have been hit by hurricanes in recent years. In 1992, Hurricane Andrew cut a swath across Miami and southern Florida, destroying more than 25,000 homes. Hurricane Gustav severely damaged Baton Rouge in 2008. And neither of those storms compared to the killer that took an estimated 8,000 people in Galveston, TX, in 1900.[22] So why did Katrina and Rita capture our long-term attention?

The answer may well lie in the way the story was told. Obviously, improvements in technology made it possible for people everywhere to see the extent of the destruction in ways they could not see in 1900—or even in 1992 during Andrew's destructive reign.

But coverage of Katrina and Rita was focused on people. Journalists paid attention to those they are supposed to serve. Furthermore, the coverage was turned into a morality play in black and white. The hurricanes spawned clear victims, the poor of New Orleans and along the Gulf Coast, and clear villains, the ineptness of government officials more concerned about their dinner reservations than about the thousands who had no dinner. As a result, correspondents shook off years of training in being bland and took on Murrow's cloak of righteous indignation.

Such coverage resulted in an audience that was engaged—and stayed engaged. Years after the flooding of New Orleans, the impact of the hurricanes remained a national story.

The stories resulting from Katrina and Rita were so dramatic they could not be hyped, so journalists came across as honest. And the public responded because the stories were real and the anger was real. At least for a short time, a difference was witnessed between fabrication and reality, between selling and reporting, campaigning and governing, expressing and existing.

But perhaps the greatest lesson to be taken from the coverage of Katrina and Rita is that today's media have leadership roles in society, whether they like it or not. They may be, as Crichton so sarcastically noted, narcissistic self-serving reporters asking thoughtless questions and thus no different from the narcissistic self-serving politicians who evade those questions. But they must accept the fact that, in today's media-dominated society, they play a significant and expanding role. The relationship between reporters and politicians must be viewed as symbiotic, a dance between near-equals that profits not only reporter and politician but the community as well.

Much has been made of the anger shown by reporters after the failure of government to respond effectively to Katrina. Clearly, that anger was critical in establishing a bond between reporter and audience.

But this does not mean that reporters and anchors must adopt anger—or even righteous indignation—as a new mantra to replace the traditional fair-and-balanced concept. The nation today is perhaps more divided than at any time since the Civil War—and a great deal of that divisiveness is the result of self-serving entertainers who use anger and employ code words to rally listeners to their cause.

Neither does it mean that reporters should present only one side to every story. That also is a disservice to society. The goal must be truth—and if truth results in righteous indignation, the anger will come off as truthful. Reporters can no longer cop out by saying, "I presented both sides of the story so my job is finished." They must become active seekers of truth, comparing and contrasting different viewpoints to find the one that best represents reality.

This also means that reporters must become far better trained than they currently are. Society cannot afford questions that presuppose answers or general questions that can be generally evaded. Reporters must become women and men for all seasons, broadly educated, questioning individuals, with strong critical thinking skills and strong historical perspectives. The profession cannot progress if it insists on hiring dullards.

If journalists adhere to their norm of presenting both sides of every issue and of taking no position themselves, they will be increasingly used by the politicians and the public relations people employed by the politicians to manipulate the media.

That is of no service to society. If media are to have any purpose other than entertainment and titillation, they must get off the sidelines and become active participants in the democratic process.

"The politics of all this are very simple," Brian Williams of NBC News said on the air. "If we come out of this crisis and in the next couple of years don't have a national conversation on the following issues—race, class, petroleum, the environment—then we, the news media, will have failed by not keeping people's feet to the fire."[23]

Notes

1. Hamilton, J.M., Burnett, J.F., Wells, K., & Thompson, I. (2006, October 28). Journalists Report: Katrina and Her Aftermath, Louisiana Book Festival Panel Discussion, Baton Rouge, LA.

2. Overholser, G. (2006). On Behalf of Journalism: A Manifesto for Change. Retrieved April 30, 2009. from http://www.annenbergpublicpolicycenter.org/Overholser/20061011_JournStudy.pdf. See also: Overholser, G. (2008). Updating "A Manifesto for Change." Manship School of Mass Communication's Reilly Center for Media & Public Affairs. Retrieved April 30, 2009, from http://www.genevaoverholser.com/?q=node/14.

3. Williams, Brian (2006, August 28). Katrina, the Long Road Back, NBC News Special.

4. Haynes, M. (2006, December 22). A Tragedy's Emotional Impact Can Engulf Journalists.

5. Haynes, M. (2006, December 22). A Tragedy's Emotional Impact Can Engulf Journalists, Pittsburgh Post-Gazette, Pg. E1.

6. Cooper, A. (2005, September 1). 360 Special Edition, CNN.

7. Dart Center for Journalism and Trauma (2005, September). A Sense of Outrage: Covering Katrina's Aftermath. Retrieved September 8, 2007, from http://www.dartcenter.org/articles/personal_stories/hewitt_gavin.html.

8. McFadden, K. (2005, December 30). Emotional surrender; The Sci-Fi Invasion Fizzled While TV On-Demand Soared, But Passion Was Everywhere in 2005, From Shows Skewed To Women To Coverage of Hurricane Katrina, Seattle Times, Pg I1.

9. Overholser, G. (2006). On Behalf of Journalism: A Manifesto for Change.

10. Deggans, E. (2005, September 8). Journalists' Outrage Visible in Coverage, St. Petersburg Times, Pg. 6a.

11. Overholser, G. (2006). On Behalf of Journalism: A Manifesto for Change.

12. Chin, O. S. (2006, January 7). In 2005, Journalists Became the News.

13. Cooper, A. (2006, June 1). Larry King Show, CNN Transcripts. Retrieved April 13, 2008, from www.cnn.com/transcripts.

14. Chin, O. S. (2006, January 7). In 2005, Journalists Became the News, The Straits Times, Singapore. Retrieved on September 8, 2006, from http://www.asiamedia.ucla.edu/article-world.asp?parentid=36786.

15. Gabler, N. (2005, October 9). Good Night, and the Good Fight, The *New York Times*, Pg. 12.

16. Crichton, Michael (1993, April 7), Mediasaurus: The Decline of Conventional Media, speech to the National Press Club, Washington, D.C. Retrieved April 22, 2009, from http://www.crichton-official.com/speech-mediasaurus.html.

17. Crichton, Michael. (1993), Mediasaurus: The Decline of Conventional Media, speech to the National Press Club, Washington, D.C., April 7, 1993. Retrieved April 22, 2009, from http://www.crichton-official.com/speech-mediasaurus.html.

18. Cough, Paul J. (2007), The case of the disappearing TV Viewers, Reuters, May 25, 2007. Retrieved April 22, 2009, from http://www.reuters.com/article/entertainmentNews/idUSN2523545420070525.

19. Goldberg, J. (2007, September 5). Storm of Malpractice: Katrina was a media disaster. National Review Online. Retrieved September 8, 2007, from http://article.nationalreview.com/print/?q=NmEyNj MzMWQ1OTI3ZjhiMmE5YWNkZDc-2MmM2NDQ1NTg=.

20. Adams, C. (undated). In the wake of Katrina: Going home to a "no-man's land," (brochure). American University School of Communications. Retrieved September 8, 2007, from http://147.9.1.138/main.cfm?pageid=1365.

21. de Zengotita, T. (2002, April). The Numbing of the American Mind: Culture as Anesthetic, Harper's Magazine, Retrieved September 8, 2007, from http://www. csubak.edu/~mault/Numbing% 20of%20american%20mind.htm.

22. NOAA Technical Memorandum NWS TPC-1 (1997, February). The Deadliest Hurricanes in the United States, 1900-1996. Retrieved September 8, 2007, from http://www.nhc.noaa.gov/pastdead.html.

23. Williams, Brian. (2006, August 28). Katrina, the Long Road Back, NBC News Special.

2

Hurricane Katrina:
Flooding, Muck, and Human Misery

Guido H. Stempel III

Darkening skies in New Orleans foreshadowed the impending arrival of Hurricane Katrina. For days, weather satellites monitored its size and strength. Meteorologists plotted its every twist and turn. Civil authorities and the media issued repeated warnings for citizens to evacuate. Unlike 9/11, there was no unseen enemy—just Mother Nature in a foul mood.

That foul mood was reflected by meteorologist Robert Ricks when he issued an urgent message from the National Weather Service: "All gabled roofs will fail. All wood-framed low-rising apartment buildings will be destroyed. All windows will be blown out. The vast majority of native trees will be snapped or uprooted. Only the hardiest will remain standing, but will be totally defoliated. Livestock left to the hurricane winds will be killed. And finally, water shortages will make human suffering incredible by modern standards."[1]

There was no room for surprise about Hurricane Katrina and its ultimate companion Hurricane Rita. The news media had several days to marshal resources, move camera crews, and prepare coverage of the initial storm. The eventual surprise came not in a quick series of spectacular 9/11-style explosions but in the deadly inches—feet—of floodwater that accumulated throughout the day.

Hurricane Katrina attacked Louisiana, Mississippi, and Alabama on Aug. 29, 2005, and was followed less than a month later by Rita's storming of Louisiana and Texas. Katrina was the biggest natural disaster in the history of the United States. More than 1,800 people were killed, and 33,544 were rescued by the Coast Guard. Hundreds of thousands were made homeless, at least temporarily. Evacuees resettled in all 50 states.[2]

Estimates vary, but economic loss through property damage exceeded $81 billion. Energy industries especially were hard hit, with the loss of 10 percent of the annual production of petroleum and 8 percent of the annual production of natural gas from the Gulf of Mexico.

It was a story made for television, and coverage featured strong visual images of the massive destructive force of two Gulf Coast hurricanes and stories of its impact on the residents of the area, said John Kieswetter, TV/media writer for the *Cincinnati Enquirer*.

"The raw as-it's-happening live reporting allows everyone to know the latest information about a crisis at the same time," he said. "In a time of crisis, live TV coverage is the greatest resource for American society—the free flow of information from competing news organizations trying their best to shed light and provide perspective on a complex situation."

And the best that television—indeed, all news media—have to offer, according to the National Research Council Committee on Disasters and the Mass Media, is to fulfill six specific functions: warning, providing accurate information to the public and officials, charting the progress of relief efforts, dramatizing lessons learned, taking part in long-term education programs, and analyzing longer term issues that result.[3]

This is the context for this study that examines coverage of Katrina by the five television networks—ABC, CBS, NBC, CNN, and FOX—for the first 24 hours of the hurricane.[4] Unlike the coverage of 9/11, the networks did not suspend their regular programming, so this study focuses on their regular news presentations.

Coverage on the Scene and from the Studios

What the networks did, of course, was provide massive coverage by reporters and camera crews on the scene and anchors in the studios. It is not surprising, given its 24-hour operation, that CNN had far more coverage than the other networks, about three times as many stories as ABC, CBS, and NBC and 50 percent more than FOX. Overall, CNN gave the hurricane five times as much time in seconds as ABC, CBS, and NBC and nearly three times as much as FOX. The average story on CNN (3.5 minutes) was nearly twice as long as the average for the other four networks (less than 2 minutes).

Katrina itself was not selective in its attack, pounding the residents of three states with ferocity, but the networks gave the largest amount of their attention to New Orleans in the storm's first 24 hours. NBC's coverage led the way in this regard with nearly 62 percent. No other network exceeded one-third. Reporting from Mississippi comprised 18 percent of the total, and Alabama received little attention, with CBS and NBC having none at all. FOX, on the other hand, had nothing from Louisiana outside New Orleans.

All networks had journalists on the scene, but a great deal of the coverage was simply from the studios. Considerable variation existed in this regard. Overall, nine-tenths of NBC's coverage was on the scene, including the three categories of spot reporter, spot scene, and spot interviews. On the other hand, FOX had more stories from a reporter on the scene.

ABC, CNN, and FOX had about one-third of coverage from the studios, and CBS had slightly more than one-fourth from the studio. FOX had no interviews at all, but it is not surprising given that in the first 24 hours interviews were relatively scarce on all networks.

In spite of all the warnings, the amount of coverage from the studio shows how unprepared the networks were for a disaster the size of Katrina. Hurricane coverage is rather basic normally. Camera crews set up in safe locations and wait for the storm to pass before venturing out to gather film. The biggest difficulty usually is downed power lines or partially submerged streets.

But Katrina produced few partially submerged streets in New Orleans. Instead, journalists and the public had water deep enough in which to drown, and, in some of the suburban areas, muck deep enough to suffocate under.

Complications: Katrina Was Not a Normal Hurricane

The logistical problems, especially in New Orleans, made coverage dangerous and complicated, even when news organizations had a general plan. CNN's John Roberts, who worked for CBS at the time, arrived in New Orleans a day before Katrina hit the city. Even though he arrived before the storm and was somewhat familiar with the city, he admitted it was very difficult to get his bearings once the water began to rise.

"We didn't plan for that," he said, when asked if CBS had a disaster coverage plan. "We had plans for hurricanes based on our previous experience, and that is, have a couple of satellite trucks, park them in a safe location or a location where you think it is going to be safe from the wind, so that we can broadcast. Bring down some extra supplies with you, that sort of thing, flashlights, some extra batteries, and a couple of days worth of food just in case you can't get it and some extra cans of gas, yada, yada, yada." No one could have been prepared for what happened in New Orleans, Roberts said.

The tendency, then, to toss coverage back to the studio, was not the result of a lack of planning. It was the inevitable result of trying to cover a scene that posed nearly impossible logistical difficulties and a disaster that spanned three states. The geographic dispersion—and the difficulty in getting cameras and journalists into some areas—influenced coverage and dictated that the studio become the "anchor" for telling the story. Even local reporters, who knew the geography of New Orleans, struggled to gain an overview of the story.

"Where you were was kind of what you knew at the time," said Tim Morris, state/political editor for the New Orleans *Times-Picayune*. City editor David Meeks was more graphic. "It was like being in a hole and writing notes down on what you see in front of your face, and handing the note up out of the hole, and someone takes the note out of your hand and you don't ever know what happens to it."

The lack of basic amenities also contributed to the problems of coverage. Roberts, for example, started off his coverage for CBS while living in a hotel room on high ground but was forced out of his hotel after the storm passed. He spent one night sleeping on Interstate 10 and another sleeping on top of his car, which was parked across the street from Harrah's Casino near the French Quarter.

Even Ted Jackson, a *Times-Picayune* photographer with an intimate knowledge of the city, had difficulty finding a place to sleep. "It was so hot, you usually slept outside on the porch with the mosquitoes. That was awful," he said.

> And one night we got word from a friend that we could sleep at her house but we would have to break in because there was no key. And we broke into the house and, as I went through the house to open the front door for the other photographer, I met a shotgun raised to my head. He was a retired New Orleans police officer intent on protecting his neighborhood. He thought we were looters. That was a tense moment. He had been told by deputies in the neighborhood that 'If you see looters, shoot them and throw their bodies in the canal. Don't ask any questions.' That was exactly what he was intending to do.

The experience of Manuel Torres, a *Times-Picayune* reporter who covered St. Bernard Parish at the time, illustrates the problems reporters faced in getting into some regions.

"St. Bernard had a different kind of flooding than Orleans," said Torres, now a *Times-Picayune* assistant city editor. "In Orleans, there was a lot of debris and muck in the immediate areas around the breaches especially the middle part of New Orleans, northeastern New Orleans. In St. Bernard, the main reason for flooding was the surge that overtopped the levees and basically dredged the marshes and dumped all of that sediment and grass on the parish. So you had about two feet of muck that covered most of the parish. So even after the water receded it wasn't just a thin layer of mud that had remained—it was really impassable."

But the local journalists did have some advantages. They understood the topography of New Orleans and the surrounding area. They knew where problems were likely to occur, so many, print and broadcast, were pre-positioned around the city and throughout the region in advance of the hurricane, assigned by their stations or newspapers to cover the region they normally cover.

The network journalists, on the other hand, tended to congregate in central locations, said David Meeks, city editor of the *Times-Picayune*. "So they came in, and they tended to set up their stations on Canal Street in front of the Sheraton Hotel, near the police operations and on top of the raised elevated interstate, aiming their cameras either at the Superdome or the Convention Center. That was a limited corridor from which they did their reporting."

Even though the perspective might have been limited, the national journalists did what tradition indicates they are supposed to do because their responsibilities are different and for a different audience from those of their local counterparts, said Chris Slaughter, news director of WWL-TV in New Orleans. "Local television's job in a crisis like Katrina is to inform the people, to provide the most complete information we can to help them make decisions that are important to them, for example, whether and when to leave the city. People need to know where to go for help, what has been the impact in what neighborhood. We worked hard getting information, making it available to the people clearly, accurately and rapidly through various platforms, on air, Internet, through radio."

The national networks, on the other hand, focus on a broader perspective, he said.

> Someone in Kansas doesn't need to know when or where to evacuate or what New Orleans neighborhoods have the greatest problems. The nationals focus on the bigger picture. They provide information on what happened, sure, but they deal with stories from abroad, about government, about problems in the future.
>
> They don't know the lay of the land, so a lot of them work with us. They're not as good at getting local information, but they may have access to the head of the National Guard or FEMA. But we know the cop on the street. So material flows both ways. We meshed well together.

Television Coverage Focused on Facts and on Victims

Despite these limitations, national television reporters gave powerful images of the destruction and human degradation that became the stories of Hurricane Katrina, and they did so with concentration on fact. All networks but one concentrated 88 to 100 percent (CNN) of their coverage for the first 24 hours on factual information.

Fifty-six percent of FOX's coverage was factual and 25 percent was analysis, which was far more than was provided by the other networks. In the first 24 hours of the storm, little consolation or guidance was provided, which probably reflects the reality that so much factual information was readily available and extremely important. This changed as the week progressed.

In addition, these national journalists—sharing in at least some of the danger and physical suffering of New Orleans residents, provided coverage that also was characterized by compassion. All networks except ABC focused on compassion more than any other factor in the first 24 hours of coverage. ABC emphasized human rights more than compassion, but CBS, NBC, and FOX emphasized compassion over human rights by a ratio at least 5 to 1. Only CNN emphasized competency to an appreciable degree.

In the context of the first 24 hours of Katrina, it was clearly appropriate that the most frequent frame of the coverage was disaster, with an average for the five networks of nearly 60 percent. Economy was a strong second on CBS, but far lower on the other four networks. Safety was second overall, but not on NBC where the environment was second.

Much of the coverage went beyond the physical damage to the city and focused explicitly on its emotional impact. Local journalists, after all, were victims themselves, and they shared strong feelings with other residents about what they saw happening.

"It was very much like 9/11," said Dan Shea, managing editor for news at the New Orleans *Times-Picayune*. "I remember when the first tower came down, I was watching MSNBC and even though you saw it with your eyes, they were reluctant to say it had just come down. They said 'something is going on, part of the building may have fallen, part of it may have fallen off. We are not sure.' To sit there and to say the city you live in and love is going to be in large part destroyed is difficult."

Those with the networks demonstrated that they sensed, and perhaps even shared, the emotional turmoil in New Orleans. In most respects, what was happening to the people was the story that needed to be told, and this was reflected in the overwhelming concentration of the coverage on victims.

The victims, by far and away, comprised the leading topic of the discussion. Nearly three-fourths of the stories were about victims, and the range was from 64.7 percent on NBC to 87.5 percent on ABC.

The environment and FEMA were next. Surprisingly, the mayor of New Orleans was the subject only on FOX and not very much there.

Victims also represented the most frequent issue discussed on the networks followed by severity of the storm. Severity was No. 1 for ABC and FOX. Economy was second for CBS, but was barely touched on by ABC, CNN, and FOX. CNN covered rescues considerably more than the other networks.

Furthermore, victims were the major sources. Rather obviously, victims were available in large numbers from the beginning. Various experts were the second most common source, followed by Louisiana Governor Kathleen Babineaux Blanco. Surprisingly, the National Weather Service was not the source for any stories on ABC, CBS, and FOX, and it was not a major source for the other two networks.

These findings illustrate various differences in coverage as well as what the networks had in common. The results do indicate that the networks were not as well prepared as might have been expected, given that the time of the hurricane was predicted with reasonable accuracy. This is reflected in the relatively high percentage of stories that were broadcast from studios rather than the sites. A striking difference emerged between NBC and the other networks in the number of stories coming from New Orleans.

Another striking result is that President Bush represented 30 percent of the total sources for FOX, but was not a source at all on ABC and CBS and a source very little on NBC and CNN. Because he was on a political speaking engagement in California, Bush was not readily available for much of the first day of the hurricane.

Also missing during the first 24 hours is much indication of activity of FEMA, but that was a reflection on FEMA, not on the media.

Katrina and September 11

The most obvious difference between the coverage of 9/11 in the first 24 hours and the coverage of Katrina in the first 24 hours is that the networks provided far more stories about 9/11 than about Katrina. September 11 coverage resulted in 2,647 stories during the first 24 hours compared to 133 about Katrina, a ratio of 20 to 1. This reflects in part the fact that all networks provided all-day coverage for 9/11. On the other hand, the broadcast networks, ABC,

CBS, and NBC, did not alter their daily programming because of Katrina—or at least did so on a very limited basis. It is possible that they had updates that were inserted between programs and did not become part of the Vanderbilt archive on which this study is based.

However, the difference in number of stories also indicates a difference in the greater extent to which 9/11 dominated coverage. Katrina was the major story of its day, but 9/11 was close to being the only story. The 9/11 coverage was rather evenly divided among the networks except that FOX was considerably lower. However, with Katrina, FOX had about twice as many stories as ABC, CBS, and NBC, and CNN had nearly three times as many.

Heavy emphasis was placed on disaster in the coverage of both 9/11 and Katrina, but 9/11 involved considerable political coverage compared with virtually none for Katrina during the first 24 hours. The political implications of 9/11 were clear from the beginning. Political aspects of the Katrina story emerged later, being more associated with the governmental response to the storm. In both cases, description of the event was the major part of coverage, but it took on a different character with Katrina as the hurricane was still going on.

Physical Damage, Human Suffering, and Political Meaning

Circumstances, rather than policies or even preferences, often dictate coverage provided by the media in a rapidly moving and complicated crisis situation. Thus, it was circumstances, with some dashes of policy and preference, that produced differences in the coverage of 9/11 and Hurricane Katrina. Both stories had startling images, massive property damage, devastating impact on people, and powerful political implications. But the major question during the first 24 hours of coverage of each tragedy was what information was most readily available.

On that fateful 2001 day in New York, Washington, and the isolated field in Pennsylvania, what was most immediately available initially were images of the property damages, airplanes striking the twin towers, smoke billowing from the buildings, the unbelievable collapses, and heaps of rubble in three separate locations. The circumstances certainly led journalists immediately into speculation about human loss, but they had few images and little information about human beings. So, as communication theory suggests, their initial focus was on the facts, on trying to explain what happened.

The surprise of the 9/11 attack and its obvious worldwide political implications prompted journalists to begin their search for who was responsible immediately. In so doing, they sought out government officials and foreign-policy experts to offer explanations although even the officials and the experts knew little in those early hours. Initially, victims took a back seat, although television later began to assess the human damage and personal stories.

For Katrina, however, images of the storm itself and even the flooding were superseded quickly by stories about real people who were suffering. Such images of victims were readily apparent and compelling. Journalists tended to talk with whomever was available. Victims were plentiful and outspoken. Officials were not, and even those who could be reached had little of significance to say. Katrina and Rita did produce visuals about property damage, but both hurricanes were human stories—especially in New Orleans—from the beginning, a fact reflected by the networks' dominant focus on victims during the first 24 hours of coverage.

How journalists dealt with the political implications of the disasters likewise was different. The shock of the unexpected 9/11 attacks and the compelling pictures of the smoking, and ultimately collapsing, twin towers resulted in a personal journalistic numbness that was overcome only through journalistic instinct, including the will to determine who was responsible. On the other hand, when the levees failed in New Orleans, the media concentrated on the immediate impact. Analysis of who was at fault came later, so the politicians and the U.S. Corps of Engineers got a break initially only to get the full force of public and journalistic anger later.

Over time, the networks balanced their coverage. They told of the human suffering, the physical damage, and the broader implications. And in both cases, the best among the media continued the stories for years to come. Neither 9/11 nor the Gulf Coast hurricanes have reached the end of their importance to media audiences—or to journalism.

Notes

1. Williams, Brian. (2006, August 28). We Were Witnesses, NBC News. Retrieved September 9, 2007, from http://rss.msnbc.msn.com/id/14518359/; Williams, B (2005, Sept. 15). The Weatherman Nobody Heard, NBC News. Retrieved September 9, 2007, from http://www.msnbc.msn.com/id/9358447/.
2. United States Congress. (2006, February 19). *A failure of initiative: Final report of the select bipartisan committee to investigate the preparation for and response to Hurricane Katrina.* Washington, D.C.: Government Printing Office.
3. National Research Council Committee on Disasters and the Mass Media. (1980). *Disasters and the mass media: Proceedings of the committee on disasters and the mass media workshop.* Washington, D.C.: National Academy of Sciences.
4. This study was conducted with videotapes from the Vanderbilt News Archive. The length of stories was recorded in seconds, using time notations placed by the archive on the tapes.

3

NBC News: Covering a Tale
of Human Suffering

Ralph Izard

*Katrina had left an indelible mark on Brian Williams. It was the story that
defined him as a network anchor, the story that he was determined to own, the story
that haunted him in ways that were impossible to explain to anyone who had not
seen the devastation firsthand. After making several trips to New Orleans, he had
insisted that his wife, Jane, come along. "If you're going to live with me," Williams
told her, "you're going to live with this story."*
—Howard Kurtz, the Washington Post [1]

Brian Williams, anchor and managing editor of *NBC Nightly
News*, was at home on a Saturday night (August 27, 2005) as his
family prepared for a summer vacation. He received a call from
Steve Capus, president of NBC News, who had news about Hurricane Katrina, rapidly approaching the Gulf Coast, possibly with
special malice for New Orleans.

More than a year later, Williams remembered vividly the urgency
in Capus' voice: "This is a monster. I've just gotten off a conference
call with Max Mayfield (director of NOAA's Tropical Prediction
Center at the National Hurricane Center in Miami). It appears to
be the 'doomsday scenario' ('He used those two words,' Williams
recalled), and I think we've got to go."

Williams was surprised. A self-described "weather hound" who
keeps the weather channel on one of the six monitors in his office at
30 Rockefeller Plaza in New York, said because of the storm's history he had not locked into Katrina as firmly as he might have.

"It had fluctuated so wildly, and we thought that the 5 status was a kind of sudden, almost perverse, overcharged microburst, and that it was so large and the pressure gradient at the center was so low, we almost figured it would have to explode, collapse in on itself, and would be downgraded," he said. "Well, it didn't happen."

Williams knew he had to agree with Capus, but he added a proviso: "If we're going, my condition is, I've got to get in that Superdome. And the pieces were put in place to get me in the Superdome."

NBC chartered a plane to Baton Rouge, Louisiana's capital city about 80 miles from New Orleans. "I have a hurricane kit that I bring. I'm a former fireman, so the boots that I bought in 1976 are my pride and joy. I wear them in every storm. Every photo of me in the water in Katrina shows me in them."

The chartered Learjet landed in Baton Rouge on Sunday as the storm was bearing down on the Gulf Coast and after a wild, bumpy ride. Williams and the NBC crew went to a Wal-Mart to add to their supplies.

"The Wal-Mart was being looted in a civil fashion," he said. "People were paying for goods, but the shelves were being looted. People were grabbing anything. Anything edible—it looked fantastic. Vanilla pudding—24 pack—fine—we can live off that. The kind of milk-size gallon jugs of no-name water that taste of the gallon jug—we bought some of those—any port in a storm. Probably consuming as many carcinogens as we would in a lifetime."

Getting lodging was difficult in New Orleans, a city that was shutting down in the path of a storm that could be the most powerful, most dangerous, in the history of a city accustomed to dealing with hurricanes. But arrangements were made for lodging at the Ritz Carlton, "of all places," on Canal Street at the edge of the world-famous French Quarter, now eerily deserted.

A Tale of Human Suffering and Degradation

The NBC staff went to the Superdome, which would eventually to be populated, the New Orleans *Times-Picayune* reported, by as many as 26,000 people. Mostly poor and mostly African Americans, these folks had either lost their opportunity or had no means of escape; they cowed before the storm in what turned out to be a futile hope that they would be safe.

"We settled in for the storm," Williams said. "The corrugated steel doors were closed," and NBC crew members began reporting what they witnessed. "I tell of the people I met there and the winds and the rumors inside, the suicide we witnessed. A guy jumped off the upper deck. At times you couldn't find a member of the National Guard. They were hunkering down. The atmosphere became violent, the air was dank, nothing circulated, no power, of course, restroom facilities were knocked out after the first night. People in my mind and in my view (something I've said a lot) displaying great dignity in horrible conditions, and I find it deplorable and heartbreaking when I think back on that memory." When time came to do that evening's "Nightly News," Williams and his crew were allowed to leave the Superdome. They did their evening show from the Sports Complex adjacent to the Superdome.

"After our exit," he said, "the corrugated steel doors were closed. God forgive that people should be allowed to leave the Superdome. I even paused on the air and said about the people we had just left behind, 'There's no air, food, or water, and no instructions. Perhaps worst of all, no instructions. They don't know what's happening outside, or what's going to happen to them.'"

Williams' struggled to do what needed to be done, partly because of the intense emotion he was feeling as he witnessed human degradation and partly because he was suffering from dysentery. "I have little or no memory of long stretches of that week because I was very ill. They (NBC) were a little worried that their anchorman was going to pass out on live television. I was completely dehydrated, running a fever, had no electrolytes in my body. They found me some sports drink and sat me on an equipment case, and I was somehow able to talk for half an hour in lucid fashion. Well, I guess that's in the eye of the beholder."

Williams spent a week in New Orleans reporting the impact of what turned out to be the worst natural disaster in the history of this country. During and after the storm, he was an eyewitness to mind-boggling scenes of devastation and human degradation.

I saw women menstruating. I saw adults who on Sunday had their dignity, and by Wednesday, had been forced to defecate on themselves. I saw babies reusing diapers. I saw people, who I would otherwise say "hello" to on the streets, looting stores days later. I was sympathetic to the looters. Because cash was king in a city of no food, water, transportation, or cool air. Cash could get you out of town—in a town where every tire was flat because of the shards of debris and broken glass. If you had working tires on your car, you were King of New Orleans. You heard every car go by on four rims. It was paralysis; it was the degradation of human life. I've never seen anything like it in my country.

New Orleans Did Not "Dodge the Bullet"

Once the physical storm had passed to the northeast of New Orleans, the NBC staff—indeed, other journalists, residents, and the country as a whole—breathed a collective sigh, perhaps not of relief but of the belief that "the big one" had not occurred. Damage was extensive. Windows were blown out, roofs ripped from their foundations, some buildings collapsed, automobiles strewn about like so much insignificant debris, and utility wires had ripped away and dangerously flapped in the continuing winds. But, after all, it was wind damage. New Orleans, "the bowl," had not flooded. It was serious, but it compared to hurricanes of the past.

Many journalists, including NBC News, proclaimed: "New Orleans Dodges the Bullet." "We used the expression that New Orleans had dodged a bullet," Williams said.

I'll never forget that—on that broadcast, because the streets, while wet from rainwater, were dry, in a sense that New Orleans had stayed dry, rather a miracle, considering the fact that it's referred to as "the bowl." So we could see the Hyatt, we could see that all the facing windows had been blown out. The city had taken a good blow but the city was not flooded. We could not know what was happening at that very moment at several of the levees in New Orleans.

I woke up Tuesday morning—very tired, for a *Today* show report having not slept the night before. It was scary, even in a luxury hotel, without power. We would fear fire. I put my belongings in suitcases and lived out of those. I repacked in case I needed to make a hasty retreat. Had to sleep with the window open; there were street noises, police cruisers up and down all night.

I woke up in the dark at 5 a.m. to look out at shards of light on the street. I couldn't discern, really, what it was, but the best I can explain it was like—it was as if the reflected light from a disco ball was shining on Canal Street. And then I realized I was looking at water. Something had happened overnight. Our story had changed.

"Shouldn't the Cavalry Be Coming in by Now?"

Given what Brian Williams saw during that week and in thirteen other visits to the Gulf Coast over the next two years, especially when he freely admits to being a passionate person, it should not be surprising that he frequently expressed his frustration on the air.

"I was an eyewitness," he said. "And so, any moral authority that viewers might have seen, any palpable anger that they might have seen from us owes to that, I believe." And he applied that anger, that frustration, to his reporting of the governmental failure to get immediate help into an area so thoroughly devastated and with thousands of suffering, demoralized people.

This frustration was echoed on the air by correspondent Carl Quintanilla, who said: "It's because our company was able to move so much into this city that made us ask 'Why can't the federal government do the same?'" Williams said his feelings were heightened because of his experience in Iraq where he saw the Third Infantry division land a pallet of MRE's, porta-johns, bottled water, anywhere they wished ten minutes after an order is given. "What about these people in front of the (New Orleans) Convention Center—they didn't deserve that?" he asked.

Thus, in addition to providing the details of what was happening all along the Gulf Coast and of the unbelievable suffering of human beings, Williams said, the story about governmental inaction quickly became a prime focus. It was not easy to explain to those in every American household why help failed to arrive. Thus, he and many other journalists, most conspicuously Anderson Cooper of CNN, began firing direct and accusatory questions at government officials.

"Isn't the government seeing this?" he asked. "Aren't they watching the same thing we are? Who among us didn't want to throttle our government officials, and ask them—'You say help is on the way—where is it? Shouldn't the cavalry be coming by now?' This story throttled the United States."

For his and others' questions about a conspicuously failing governmental apparatus, Williams has no apology. "What sets this apart from, say, the last time the media behaved poorly: The run-up to the war (in Iraq)? We weren't with the WMD inspectors. People like Judy Miller (of the *New York Times*) will have to answer for themselves. But (at that time) we reported a story as a broken nation, post 9/11, yearning for something from its president, and journalists got the build-up to the war (in Iraq) wrong, largely, in my view. We were witnesses to Katrina. I beat the first responders to Katrina. I was an eyewitness."

On the ground, Williams said, official government repeatedly demonstrated lack of respect for the Gulf Coast residents in their struggle for survival. He was frightened at the way the National Guard was treating people on the way into the Superdome: "very aggressive pat downs, loud instructions, yelling at the people like they have done something wrong. It was a very aggressive stance. I never did find the commander to complain to. I was told 'this is a Homeland Security show.'"

These failures, continued mistakes, and inaction over the coming months, Williams agreed, contributed to the continuing decline of the national approval ratings of the Bush Administration and to the results of the national elections of 2006 that gave Democrats control of both houses of Congress.

"I think it's a component of what we just saw the voters say," he said. "I think, to be honest, the Iraq war supersedes Katrina, if you ask voters their importance. But Katrina contributed to the patina that this administration cannot now wash off. It will always be an event that happened on their watch."

9/11 and Katrina

The governmental response over time to the terrorist attacks of September 11 and the natural attacks on the Gulf Coast thus join these two catastrophic events together in their impact on the nation. They provided emotional, intellectual, and physical challenges to Williams and other journalists from all media. Both disasters went beyond anything that may be defined as normal. Katrina and Rita were predicted, but the ultimate story, as with the September 11 attacks, was at least an immediate surprise.

In one sense, the stories were similar, but they differed fundamentally. "Both were acts visited upon us," Williams said,

> but with Katrina, it wasn't simply the storm. It was the response (that) injured people and took innocent human life. With 9/11, it was the initial act that injured people and took human life. An act of evil vs. an act of nature—where the initial act of evil was the killer, as opposed to benign neglect and real wrongdoing on the part of the federal government.
>
> (Former New York Senator) Daniel Patrick Moynihan is somewhere in legislative heaven, and he is looking down, I hope, with mixed emotions on how many times his phrase intended to have one use, one meaning, has been used and re-used, I hope with due credit. He described so many things in American life, but "benign neglect" lives on as the perfect descriptor of so much of what we have witnessed.

No Journalistic Short Attention Span This Time

For these reasons, Williams and others have pledged that the ultimate story of the hurricanes will not fall victim to the normal journalistic short attention span. It's common practice, when an incident occurs, for national journalists to "parachute" in if the location is outside their normal haunts. They pay extensive—sometimes overwhelming—attention for a few days, then leave for the next assignment. What was a major incident or tragedy is pushed into the background, perhaps even forgotten.

This will not happen with the efforts on the Gulf Coast to rebuild following the devastation of hurricanes Katrina and Rita, Williams said. It's something to which Williams is dedicated. As Howard Kurtz of the *Washington Post* said later, the NBC anchor believed that without sustained attention from the national media, New Orleans would never recover.[2] And Williams credits his boss, Steve Capus, president of NBC News, who recognized that the Gulf Coast story "is going to be one of the stories of our times."

Thus, two months later, NBC opened a new bureau in New Orleans that continues to provide coverage of the aftermath of the storms, supplemented by national correspondents who flow in and out. Williams himself went to the Gulf Coast six times in 2005, five times in 2006, and two times in the first half of 2007.

"A million people displaced, an area of land affected that's larger than the United Kingdom, a huge class angle to this, a huge race angle to this," he said.

We can't just wake up tomorrow morning and say that everything is fine.

So NBC News has decided that this region is worth our investment. And it is too important to the United States. Pick an area—energy, culture, employment—we will still go back and do Thanksgiving among the refugees. This has redrawn the political map—this changed our voting trends, this made accurate polling impossible for the mid-terms, for the mayoral race in New Orleans—you name it."

An Award-Winning Team

The work of the NBC News team did not go unnoticed. That Williams and his colleagues were on hand when the storm hit and continued their coverage was recognized officially and repeatedly. The extensive coverage of the story as it occurred and the network's frequent special programming earned a Peabody Award, an Alfred I. duPont-Columbia University Award, an Emmy Award, three Headliner Awards, three Edward R. Murrow Awards from the Radio-Television News Directors Association, and a Sigma Delta Chi Award from the Society of Professional Journalists.

Williams is proud of this recognition. And while he knows he worked very hard to provide the American people with good coverage of the impact of hurricanes Katrina and Rita, he admits to "a lot of guilt of being the front man." He knows his network's award-winning coverage was the product of many people on the NBC team. He praised not only news correspondents Lester Holt, Martin Savidge, Don Teague, Carl Quintanilla, Campbell Brown, Kerry Sanders, and Ron Mott who appeared on the air or the executives who provided guidance and financial resources, but others whose efforts made the coverage possible.

"A guy named Walter Parks who drove our camper in to the middle of the city, but first stopped at a Costco and tried to guess at everything we would want to eat, because he was coming into a city of dirty, tired, and hungry people. And he knows today how I feel about him. Tony Zumbado is a cameraman whose personal testimony was so stark and elegant, and he tells on live television what he saw inside the Convention Center. So that's got to be stressed, with all of the great rewards and recognition we received in covering Katrina, my frustration is they all can't be onstage at one time."

Notes

1. Kurtz, Howard. (2007). *Reality Show: Inside the Last Great Television News War* (pp. 170-171). New York: Free Press.
2. Kurtz, Howard. *Reality Show: Inside the Last Great Television News War,* (p. 171).

4

Local Coverage: Anticipating the Needs of Readers

Roxanne K. Dill*

Veteran reporter Anita Lee had experienced enough Mississippi Gulf Coast hurricanes to know when it was safe to venture
outside. She and some of her fellow staffers waited out Hurricane
Katrina in the Biloxi Sun Herald building, a concrete bunker built
nearly forty years before when Hurricane Camille destroyed the
previous one.

Lee's decision to remain in Biloxi was practical: If she evacuated, she was told she might not be able to get back in to report or
to edit. Besides, she, as did others, expected Katrina to be like so
many other storms, leaving behind wind damage and downed trees,
with business as usual as soon as power was restored.

The fact that her home was a block from the beach did not worry
Lee and her roommate Margaret Baker—also a reporter. Lee care-
fully wrapped her precious photos and put them in a trunk on top
of the bed to avoid water damage before heading to the hurricane
campout at The Sun Herald building. As long as The Sun Herald had
Internet service, she and others continued to send news updates.

But conditions quickly deteriorated. The wind was unrelenting.
Most difficult for Lee was waiting to see what the hurricane had
done. "A crazy freelance photographer staying with us in the build-

---

* This study by Roxanne K. Dill originally was conducted as a master's thesis
in the Manship School of Mass Communication, Louisiana State University.

ing insisted on going out to shoot photos. I implored him to wait because I was afraid a flying object would hit him in the head or chest and kill him," Lee said. "He went out anyway, came back, and looked like he had been in a war zone or something. He had this shocked look on his face and the first thing he said to me was, 'Your city is gone.'...That's when I knew it was bad."

When the last of Katrina's winds quieted, Lee grabbed her notebook and headed outside, climbing over trees in search of a story. "I was prepared to interview devastated people, but when I saw my house, I realized I was one (of those devastated people)." The wind destroyed her home; the storm surge took what was left inside and carried it into the Gulf of Mexico. "After I saw the house, I had to suck it up and move on, and get focused on what we were going to write."

What Lee did not know in those early hours after Katrina moved on was that the storm would provide her and fellow newspaper reporters in the field an unparalleled autonomy that would allow—indeed require—them to focus on the needs and desires of their reading audiences rather than the pull of the media pack.

Playing "Follow the Leader"

This is contrary to normal journalistic practice. Research has shown that reporters and editors are like most people—they prefer to follow the leader. Scholars have discussed in erudite terms what most armchair nightly news connoisseurs already knew: The news appears to be the same, from day to day, from channel to channel, from newspaper to news magazine, and from one radio station to the next. Media organizations tend to look to each other and, importantly, they look to elite news organizations for guidance and to reduce uncertainty about what issues and events to emphasize[1] Newsroom practices reveal a "dendritic" flow of influence through which editors of smaller papers look to editors of larger papers for guidance about how to make up the daily news page.[2] This is especially true if the original stories appeared in either the *New York Times* or the *Washington Post*.[3]

This tendency of news gatherers to emulate each other, combined with the massive need for news created by Hurricane Katrina, introduced the possibility of a perfect storm of copycat coverage and bad media behavior. Decades of research about reporting of disasters often demonstrates that what people get is a distorted, mythical, and perhaps inaccurate depiction of actual disaster behavior,[4] focusing upon the most dramatic, visual, or exciting elements to the exclusion of the most significant.[5]

Examining Newspaper Coverage at Three Levels

Enter Hurricane Katrina. What did newspapers contribute to our mental scrapbook of the nation's most disastrous natural disaster? This content analysis of 263 front-page stories presented by six newspapers during the two weeks following Katrina compared local, regional, and national coverage of the disaster. It asked what topics were being covered, what sources were being cited, how the coverage was framed—the way in which journalists present an event to achieve the maximum audience interest while simplifying and giving meaning to the event.[6]

The analysis looked at who or what were to be assigned blame for delays and the resulting evacuee distress after the storm. And especially through interviews with reporters and editors, it sought understanding of the process of covering this disaster and its impact on those who provided that coverage.

Elite newspapers like the *New York Times* and the *Washington Post* provided the national focus. Two local newspapers, the *Times-Picayune* in New Orleans, Louisiana, and the Biloxi, Mississippi *Sun Herald*, continued publication in large part because of their dedicated staffs—many of whom lost their homes in the storm—and the help of publishers outside their affected areas.

The *Times-Picayune* did not print for several days after Katrina made landfall, but digital versions of the newspaper were available for coding purposes. The paper later was printed with the help of publishers outside of New Orleans. The *Sun Herald* moved its printing operation to a sister paper in Georgia prior to the storm's landfall. Two regional papers, the *Advocate* in Baton Rouge, Louisiana and the *Clarion-Ledger* of Jackson, Mississippi were close enough to the storm's center to feel Katrina's winds but far enough away to avoid catastrophic damage.

Both papers serve their state capitals, which became command centers for rescue and relief operations. Baton Rouge and Jackson also absorbed tens of thousands of evacuees, immediately increasing population and affecting management decisions of their respective newspapers.

Anticipating the Needs of Readers

Clearly, readers in New Orleans and Biloxi needed different information from that required by readers in, for example, Spokane, Washington, where hurricanes pose no threat. Thus, for this study, it was anticipated that informational needs differed among local, regional, and national audiences, especially regarding the long- and short-term effects of the storm. Was this the case? The short answer is yes. Furthermore, like the levees surrounding New Orleans, Katrina eroded the theoretical expectation that newspapers would cull from the work of each other. Local, regional, and national newspapers—for practical as well as philosophical reasons—appeared to put the needs and expectations of their readers first without serious regard for the pressure of their media peers.

Of Local Concern

Topics most often presented by local newspapers were those directly affecting their readers. Local mass media systems consider disasters in their own community as "their" disasters.[7] As a result, it is not surprising that local framing is considerably different from that of national reporters who are not as emotionally, financially, and socially vested in the community. Death, injury, or illness; evacuee distress, rescue, and relief operations; and the threat of crime were presented locally with similar frequencies. Only half as often did local newspapers mention governmental failure and police activity; topics portraying conflict among government officials.

Race-based problems were virtually non-existent locally during the first two weeks after Katrina. While other media were busy contemplating the larger issues of Katrina, local newspapers were dealing with life-and-death situations. Local papers also most often presented stories that portrayed the uncertain future of evacuees.

Interestingly, local newspapers—which had more reason than any others to portray despair on Page One—were three times more likely than any others to present stories in a positive light. This frame selection parallels the tendency of local newspapers to present positive topics in reporting the disaster, as an encouragement to their readers as well as a testament to others who may have seen evacuees as helpless and without initiative. People along the Gulf Coast are accustomed to picking up the pieces and moving on after a storm. Perhaps their local newspapers sought to foster hope by reminding readers of their own innate resiliency while preserving one of the community's important information lifelines.

Times-Picayune Managing Editor Peter Kovacs said his staff was acutely aware of its local responsibilities. "Our main mission was to tell our readers and the people of New Orleans, most of whom weren't around here to know what was going on," he said. To accommodate a scattered readership, the newspaper posted photos of flooded New Orleans neighborhoods online. "The photos probably got more hits than anything else. It's because everyone looked at all the photos to look for something that they would recognize."[9]

The use of satellite imagery allowed viewers to find landmarks to help identify their property to determine the severity of flooding. Calls from New Orleans-area residents stranded on rooftops were entered onto the paper's web log with the location of the call, giving readers an idea of the extent of the flooding.

After the storm passed, both the *Times-Picayune* and the *Sun Herald* were distributed free to anyone still available and in need of information. Gary Raskett, circulation manager, said the *Sun Herald* was trucked 280 miles from the printing plant in Georgia to Mississippi where the papers were handed out to news-hungry residents. At one point, the *Sun Herald* had the largest press run in its history—as many as 80,000 copies—nearly double its daily pre-Katrina printing of 46,500.

"We sent reporters out with stacks of newspapers and handed them out," said Blake Kaplan, assistant city editor. "Our business is selling newspapers, but this was an extraordinary circumstance which called for extraordinary decision-making. We gave it away for the first two weeks after the storm. People were really anxious to have those newspapers. I think it has strengthened our relationship with the community. We are providing a necessary service. I guess, if anything, that I have come to view what we do as more critical to the community right now because times are so critical."

Lee agreed: "As a journalist, you're just trying to do your job, and that doesn't change because you've had a disaster. It just intensifies your motivation to do a good job. It really shows you that newspapers play a very important role in our society and in our communities. People to this day will tell you that the *Sun Herald* was their lifeline."

Regional Demand for Information

At the same time, regional papers focused on information to meet the needs of evacuees. In fact, during the two weeks after Katrina, regional papers presented informational topics twice as often as local papers and eight times more often than national newspapers. Both the *Advocate* (Baton Rouge, Louisiana) and the *Clarion-Ledger* (Jackson, Mississippi) are newspapers located in the seats of government for their respective states, which could explain why they presented news frames that emphasized the coming need for a lengthy rebuilding and reconstruction process for New Orleans and the Gulf Coast.

Waves of evacuees flooded both cities. Overnight, the population of Baton Rouge nearly doubled, making it the largest city in Louisiana. Regional newspapers felt the increased burden of providing information to evacuees, their families, and others concerned about the short- and long-term effects of the storm.

"From the outset we knew that we had an all-encompassing story because we are the seat of government where the information was flowing. We also had so many of the evacuees," said Linda Lightfoot, executive editor of the *Advocate*. "While we always covered the important events in New Orleans, we had to do much more coverage of New Orleans because many more of our readers now were from there. In those early days, we were faced, as the newspaper of record, with a broader circulation area than we've ever had."

At one point, the *Advocate's* circulation was up by an average of 10,000 to 12,000 copies per day, an increase of about 13 percent over its daily pre-Katrina circulation of approximately 95,000, Managing Editor Carl Redman explained.[13] The newspaper's Sunday circulation was up from 125,000 to 132,000 after the storm. Approximately a month after the storm, the paper printed approximately 3,000 weekday newspapers to accommodate the evacuee-fueled jump in the Baton Rouge area population. The *Advocate's* advertising space also increased as national, regional, and local advertisers, such as insurance companies, bought space to communicate with their displaced customers.

In the National Interest

National newspapers were twice as likely as local and regional newspapers to present topics that dealt with the broader impact of Katrina, for example, the short- and long-term economic effects of the storm and the government's failed response to the disaster. National papers also were more likely to discuss evacuee distress, criminal activity, conflict among government officials, and racial discrimination in rescue and relief efforts.

As events in New Orleans unfolded, several topics moved to prominence for *Post* reporters. Katrina, said *Washington Post* reporter Robert Pierre, "was one of the biggest stories in America at the time."

Post editors considered the larger question: "What would it look like if we lost an American city?" In this sense and from the vantage point of the nation's capital, a "huge human story" also needed to be told. "What was the government response? Was there a sufficient government response?" He and other reporters chose to tell the human story in a number of ways. "What did the exodus from a city look like? How was this affecting other cities across the region? How did it affect our city (Washington, D.C.)?" Katrina also was a military story, Pierre said, as the National Guard and other personnel were called to conduct search-and-rescue efforts and to help maintain order in New Orleans.

Meanwhile, as their editors deliberated these matters, he added, "We (reporters) were also just roving, just trying to figure out what was happening on the street."

Local, Regional, and National Finger-Pointing

As anxious evacuees waited for relief and an agitated public watched, finger pointing erupted. In the coverage, however, the physical storm itself—in the form of descriptive coverage—was blamed overwhelmingly at all three levels for the misery along the Gulf Coast. On the human side, however, the next level of blame was placed squarely on the shoulders of the federal government.

As the days progressed, the objects of blame increasingly took on a geographical perspective. The farther away the paper's reading audience was from the disaster, the more those quoted blamed some part of the federal government, including the Federal Emergency Management Agency, the infamous FEMA. Local papers most often blamed broken levees, state and local governments, nonprofit organizations, and, in Louisiana, Governor Kathleen Blanco for the lackluster relief response. In addition to FEMA, national papers most often blamed the Republican and Democratic parties, President Bush, Homeland Security, and a faceless bureaucracy.

National papers were twice as likely as their local counterparts to blame the evacuees themselves for many of the problems that arose in the storm's aftermath. In most cases, regional papers fell somewhere between the two, although they were far less likely than their national counterparts to blame local and state governments.

Deviating from the Agenda-Setting Path

For at least the first two weeks after Hurricane Katrina swept across the Gulf Coast, newspapers appeared to anticipate the needs of their readers rather than follow the activity of other media. The conditions resulting from Katrina's wide swath of destruction provided an environment for autonomous reporting and individualistic interpretations of events. The shared experience of reporters in the field, the lack of communication with their home offices and the rest of the outside world, and the distinct news cycle of the daily print media combined to produce an environment that did not follow the norm of traditional news coverage.

Sharing the Misery of Their Audiences

Normal journalistic practice dictates detachment. Reporters are expected to remain calm, observe what is going on, talk with sources, and inform their audiences of what they learned—not so for Katrina. Field reporters and photographers—some themselves victims—identified with the storm victims.

"We had to pull photographers and reporters (out of New Orleans) just to give them some debriefing, just to kind of calm down," said John Ballance, photo department manager for the *Advocate*.

> I don't think after covering that they will ever say "back to normal" because it changed everyone. Covering that storm changed everybody here. It was an emotional thing as well as a physically draining thing—seeing people homeless, seeing people dying, and dealing with the things they were dealing with.
>
> For the photographers who actually covered the storm it was almost like what I perceive when a photographer or a soldier in a war zone comes back with the depression or anxiety of having been in a war zone and seeing that kind of destruction. I think it's similar to something like that; I think it affects people and changes them.

The *Advocate* had to force its photographers to take time off after working sixteen-hour days, Ballance added. "And then you couldn't, because the people who were supposed to be taking off were getting in their boats and going to New Orleans to rescue people."

John DeSantis of the *New York Times,* a veteran journalist, was more personally moved by the situation he found among people who evacuated to the Superdome than any other event he has covered.

"I have never in my 20-plus years of doing this seen such human misery in such volume in one place at one time," he said. "These people came to the door of the dome as bank branch managers, supermarket cashiers, waiters, secretaries. Once you walked through that door you were no longer any of those things. Any conceivable thing that made you 'you' was gone. None of that mattered. Money didn't matter. Station didn't matter. The people were so powerless... so dehumanized. It was incredibly frightening to me as a human being. That was what struck me. I've seen all kinds of depravity, but never anything like this."

Katrina did not differentiate between newspaper employees and the general public. National, regional, and especially local newspaper reporters in New Orleans and the Mississippi Gulf Coast shared the misery of evacuees: problems with finding adequate food, water, fuel, and shelter and the ever-present communication gap.

Those who worked for the *Times-Picayune* and the *Sun Herald* shared personally in the misery and devastation of their reading audiences. In New Orleans, employees of the *Times-Picayune* joined their neighbors in the forced exodus from the city, evacuating in circulation trucks, leaving behind their homes, sometimes their families, and everything they depended upon to publish a daily newspaper.

"My most vivid memory of this was after we fled the building we went to our suburban building on the West Bank," the *Picayune's* Managing Editor Kovacs said. "We were in the parking lot, and we got together and said, 'What are we going to do now?' We were stuck there, and we couldn't get to our reporters." The answer was not simple, he added, "Every group within the newspaper had to figure out a way to reinvent what they did."

Along the Mississippi Gulf Coast, *Sun Herald* reporters were sharing in a different kind of misery produced when the winds of Katrina and the resulting storm surge leveled miles of homes and businesses. Lee and her colleagues lived in motor homes provided by Knight Ridder, which then owned the *Sun Herald*.

"One of the hardest things to deal with was the heat; there was no air conditioning anywhere, and there were no bathrooms," she said. "Conveniences you take for granted we no longer had." Knight Ridder sent in reporters and editors to help the beleaguered staff; production continued in Georgia, and papers were trucked daily to the Gulf Coast. For south Louisiana native Pierre of the *Washington Post* and other reporters, shelter was a commodity. Sometimes he drove to his parents' home several hours to the west; sometimes he was able to share a hotel room. And, like other reporters, he often slept in his car.

"I had this attitude that this was an adventure," Lee said. "We really had to put those (our losses) completely aside. When I had to report on east Biloxi, and they were pulling bodies out of the rubble, I didn't feel so bad about my losses. It was just stuff."

The *Advocate* in Baton Rouge struggled with adequate care for its photographers and reporters who were in the field with limited communication, little shelter, and dwindling food, water, and fuel supplies.

The *Picayune's* Managing Editor Kovacs admitted that his newspaper did not have a plan for how to cope with the kind of catastrophe provided by Katrina. "I think most people would say that we and every other entity in the city did not have a plan that anticipated this," he said. "We were not alone in that. The mayor and the hospitals and the police and everyone else did not have a plan. The federal government did not have a plan that anticipated this. One of the reasons I think people are sympathetic to (New Orleans Mayor) Ray Nagin when he's called on the carpet for not having a plan is that most people realize that no one had a plan."

The Curse of Communication Problems

Communication struggles were common hurdles for all reporters. With communication towers down for miles inland in Louisiana, "nobody's cell phones worked with great reliability," said the *Advocate's* Managing Editor Redman. "Text messaging seemed to be the most reliable form of communication." Some reporters were able to use satellite phones of Associated Press crews; others were able to use communication systems belonging to emergency operations in New Orleans.

Because Katrina cut such a wide swath of destruction through south and central Mississippi, communication became the *Clarion-Ledger's* biggest hurdle. "We had massive, massive problems with phones," said Debra Skipper, assistant managing editor for news and business for the Jackson, Mississippi paper.

One Gulf Coast reporter "stood under the remnants of a bridge" as the only location where she could get a signal to dictate her story, Skipper explained. "The photography department had to shoot, drive over to either Mobile (Alabama) or Hattiesburg (Mississippi) to transmit the photo." Photographers faced a ninety-minute drive each time they had to transmit a photo because of a lack of wireless capability. "They basically didn't have a choice."

The Blessing of a Communication Vacuum

The lack of good communication cultivated fertile ground for creative reporting unhindered by the agendas of other media organizations. Thus, some reporters flourished in what they do best: exercise a nose for news. DeSantis said his *New York Times* editors instructed him to go and find the stories. Pierre said he and his *Washington Post* colleagues were required to "use their own eyes and ears" to determine what was to be reported. "We were also just roving, just trying to figure out what was happening on the street," he said.

Once inside New Orleans, most reporters did not have contact with outside media sources available through television, radio, and other newspapers. "It is an interesting vacuum that you tend to operate in without the benefit of knowing what the morning's newscasts were," DeSantis explained. "What you were relying on was what you were hearing and feeling...and constructing a story from that. We didn't have a script, really."

For example, one of the first scenes DeSantis and his fellow reporter found was a gathering place along Interstate 10 in New Orleans. "We saw crowds and crowds and crowds of people standing on the interstate—people plucked off rooftops, a collection site. We didn't know the dynamic at first, so we started talking to people, finding out what their experience was. This taught us much about what was going on."

Tracking Down the Urban Legends

When they could, reporters would get some ideas from their editors and, especially, discuss the latest rumors. "If there was anything relating to stuff that could end up in print, we really, really grilled people—more than you would ordinarily grill civilians," DeSantis said. "We heard the same rumors from place to place to place. It was almost like tracking down an urban legend on the Internet."

They could not depend on some normally reliable sources, finding it was equally necessary to grill policemen, too, who were subject to rumors—rumors driven by fear and by having faced what DeSantis and other reporters described as unspeakable and unbelievable situations.

"There was such a lack of control of the situation," Pierre said. "The government did not have good information. The police chief did not have good information. All of these people were giving bad information. People would tell you about the murders that occurred in the Superdome as well as the Convention Center," he said. "That's when you found out most of the time, 'I didn't really see it. Someone else saw it.' Nobody saw all of the bodies because the bodies didn't exist. Everybody was giving bad information. I think that was one of the biggest problems everyone encountered. There were very few people who had good information."

Still, Pierre said, it was the job of reporters to deliver accurate information, to find reliable sources. "It still was the obligation of those who reported what Nagin or others said that wasn't true—to go back and check it. That's the job of the news organization."

DeSantis and his *Times* colleagues encountered the same situation. "We were in uptown New Orleans. We ran into an elderly couple who were backpacking. They said they had just left the Convention Center and they started telling us all of these horrible stories about what was going on," he recalled. "We started asking them for support for what they were telling us—'How did you know this? Did you see any babies raped?' They replied, 'My sister told me.' DeSantis said they would then ask, "Well, what did your sister see?" They would reply, "No, she didn't see it; a police officer told her," he explained. He and fellow *Times* reporters refused to file third- and fourth-hand information as fact.

Local reporters had the advantage of proximity and strong relationships built with sources. Rumors from Biloxi were filtered through to the national media but not reported by the *Sun Herald* because they could not be substantiated. Lee related one instance in which news company executives were anxious because national outlets had been reporting that forty people had been killed in one Biloxi apartment complex alone, and the *Sun Herald* had not written about it.

"Well, there was only one way to check anything you wanted to find out," Lee said. "You had to go there. So, I got in my car and went. This was a rumor. It was reported over and over in the national news, but we never reported that because we knew it wasn't true. I don't think we had a problem getting accurate information because we'd go where we needed to go."

The *Clarion-Ledger* staff, as did others, faced difficulties with corroboration of rumors because government officials were overworked and understaffed and not always available for comment. Skipper said rumors of looting also circulated along coastal Mississippi, but often simply were not true. In one such case, in Bay St. Louis, along the Mississippi coast, *The Ledger's* reporters arrived at the scene and learned that rumors of looting at one Wal-Mart were false. "In fact, the manager had basically opened the doors and said, 'Take what you need,'" Skipper said.

Information was abundant, but reliable information was hard to find. "The usual structures of authority were there, so it was not difficult to get information, but it was difficult to get accurate information," Kovacs of the *Picayune* said. At times, "rumors were the only information there was. In many cases it's like the man on his roof in Chalmette. Is he telling a rumor, or is he a truth teller? We'd quote him by name, but the normal journalistic standard would be to call the sheriff....But the sheriff was on his roof."

With a lack of available authority sources, reporters did not follow traditional methods. "I think you had to invent a model in which you quoted people saying what was going on," Kovacs said. "Some of the people said things that were so bizarre that you sort of tried to weigh it against the other evidence you had. In the case of the Convention Center...people were actually helping each other."

New Orleans Mayor Ray Nagin and Police Chief Eddie Compass were official sources who spoke of a large number of crimes occurring in the crowded Superdome and Convention Center, events which the *Times-Picayune* learned weeks later were untrue.[8] "If the mayor and police chief say something is going on, that generally opens the door for you to believe it has really happened," Kovacs said. "I feel like I understand how the Salem witch hunt happened."

Reporters at the scene began to realize that the rumors of rapes and murders in downtown New Orleans did not match what they saw. "People were sitting around there with their children," Kovacs said. "If that were happening, you'd take your family and go hide in the city." In fact, several months after the storm, the *Times-Picayune* became one of the first newspapers to document the false reports and provide accurate information. In retrospect, Kovacs said, "one of the worst things was that the false reports hindered the rescuers."

Newspaper Images of Katrina

For at least the first two weeks after the personal and physical devastation wrought by Hurricane Katrina, this cursory look at Page One content of six respected local, regional, and national newspapers indicated that images presented by the print media did not focus on the sensational content that in many ways has come to define Katrina. Both the content study and newspaper reporters interviewed indicated they were not quick to report rumor as fact.

As one editor pointed out, newspapers operate on a different cycle from that of broadcast and have more time to investigate. This gives them the opportunity to present more information than could be found in a thirty-second sound bite. Time to cogitate often prevented newspaper reporters and editors from being easily swept into media hysteria in the face of understandable public hysteria.

Without the benefit of adequate communication with each other and their home desks, reporters relied upon their instincts to find the stories. And as they worked, they shared in the misery of the main characters of their stories, most of whom did not have hot showers, hot meals, telephone access, or comfortable beds. Without the luxury of a cup of New Orleans café au lait, the *New York Times* and the morning news show reporters often did not know what they were "supposed" to be writing. So, they reported about those topics they thought were important to the folks who depended upon them.

Katrina revealed that, regardless of the perceived or anticipated content of other media outlets, newspaper journalists are invariably wedded to their audiences, especially when the need for information is greatest. And, the closer the newspaper is to the epicenter of destruction, the more it assumes the role of speaker for the voiceless thousands whose need is greatest and who have no public arena in which to be heard.

Notes

1. Reese, S. D., & Danielian, L.D. (1989). Intermedia Influence and the Drug Issue: Converging on cocaine. In P.J. Shoemaker (Ed.): *Communication campaigns about drugs: Government, media, and the public* (pp. 29-45). Hillsdale, N.J.: Lawrence Erlbaum.
2. Breed, W. (1980). *Dissertations on sociology: The newspaperman, news, and society* (pp. 195). New York: Amo Press.

3.   Gans, H. (1980). *Deciding what's news*. New York: Random House.
4.   Quarantelli, E. L., & Dynes, R. R. (1970). Property Norms and Looting: Their Patterns in Community Crises, Phyon, 31, Pg. 168-182.
5.   Wenger, D. E. (1985). *Mass media and disasters; Preliminary paper No. 98*. Newark, DE: Disaster Research Center, University of Delaware.
6.   Valkenburg, P.M., Semetko, H. A., & De Vreese, C. H. (1999). The Effects of News Frames on Readers' Thoughts and Recall, *Communication Research, 26 (6)*, Pg. 550-569.
7.   Quarantelli, E. L. (1991, January 30). *Lessons from research: Findings of mass communication system behavior in the pre, trans, and postimpact periods of disasters. Preliminary Paper #160*. Newark, DE: Disaster Research Center, University of Delaware. Written version of a more abbreviated oral presentation made at the seminar Crisis in the Media, Emergency Planning College, York, England.
8.   Thevenot, B., & Russell, G. (2005, September 26). Rumors of Death Greatly Exaggerated, *Times-Picayune*, Pg. A2.

5

Split Personalities: Journalists as Victims

A few days shy of the first anniversary of the devastating Hurricane Katrina, on August 9, 2006, John McCusker, a photographer for the New Orleans *Times-Picayune*, who had aggressively covered the storm a year earlier, made news himself. He was arrested for attempting to commit suicide.

An article that graced the metro front of the newspaper reported that McCusker had become depressed over his inability to repair his home after insurance agencies failed to process his claim.[1]

In a bizarre incident, McCusker was reported to have rammed several cars with his vehicle and attempting to pin a police officer. During the incident, he reportedly asked police to shoot him. Eventually, he was placed on six months probation, fined almost $900 and agreed to six weeks of drug testing.[2]

In an unusual show of support, friends and colleagues donated more than $9,000 to help get him through his trauma.[3]

McCusker best articulated his own downward spiral later when he spoke with Adeline Goss during a National Public Radio program called "Voices of New Orleans." "There's some nights that just in despair you lay in bed, and like you're a three-year old, you just lay there and say, I want to go home, I want to go home. And you can't go home....But then one day, maybe you get a FEMA

---

* This study by Shearon Roberts originally was conducted as a master's thesis in the Manship School of Mass Communication, Louisiana State University.

rejection letter, maybe you have a terse argument with the guy handling your SBA loan, maybe your insurance adjuster promised to meet you somewhere and he doesn't show up, you know, and anything. And you're right back to August 29."[4]

Shortly after his hearing, however, a report on NOLA.com quoted McCusker in a happier situation: "We've rebuilt our lives. I have hope now and I didn't have it then. In a way this is a great gift. I thought I'd lost everything, but I found out how many great friends I have and how much I care about them."[5]   .

Newspaper reports and public speeches have well documented how the lives of Gulf Coast journalists, such as McCusker, were dramatically altered by the storm. A 2005 *New York Times* article by Katherine Seelye, Bill Carter, and Stuart Elliot painted Gulf Coast journalists across all media as "unflinching in their commitment to the deluged city—making plain the difference between the manufacturers of widgets and the gatherers of news," despite the personal and professional obstacles they faced and continue to face.[6]

What is even more pertinent about McCusker's story, however, is how his own personal loss may have impacted his work. One of the creeds of journalism, entrenched in the profession's news-gathering routines, is the semblance of objectivity or detachment.[7] To maintain credibility with the public, journalists are trained to keep a distance from their objects of coverage and to refrain from bias or opinion in presenting the news.

But is that possible for journalists like McCusker? As a result of the storm, he became increasingly more vocal. He told a group of Brown University students who were visiting the beleaguered city exactly how angry and pent up he felt. "Katrina didn't flood New Orleans—government failure did,"[8] he told *Editor & Publisher*. Although McCusker's feelings of outrage were made public, behind the scenes lurk scores of more tales of how Gulf Coast journalists were dramatically changed by this disaster, not only in their personal lives, but professionally.

All along the coast, similar stories are told. "Seven months later, I have sunflowers growing in my yard, no house. I mean, I have no place to live right now," said Kate Bergeron, a feature writer at the *Sun Herald* in Biloxi, Mississippi,[9] where flooding was not the main force of destruction but, instead, massive wind damage that literally blew casino boats miles ashore.

In New Orleans, McCusker's colleague Kathy Anderson, a veteran photographer assigned to the Living Section, said Katrina challenged her most as a mother and a journalist. "The difficulty of covering something that you're a part of was just, you had to try to remove yourself, but at a certain point, you really couldn't." As Anderson recalled the ordeal of chronicling the demise of her city through images, she broke down in tears.

"I remember in early October going into my church on the second balcony, and I kept telling myself, that you can't really make good pictures if you're crying," she said. "I remember that, (crying), sorry, I remember that every thing that I had, for four years was gone (paused to cry). So I think for our staff and for our newspaper to do what we've done, is just an unbelievable accomplishment."

Although Gulf Coast journalists were facing the same challenges as their readers, all they had to offer was their craft. *Times-Picayune* reporter James Varney [10] recalled, "We don't have anything for these people. We don't have water, we don't have food, we don't have escape. We have nothing to give them. We were told that the main fact that the *Times-Picayune*, the local newspaper, was there was a tremendous comfort to them."

The fact that Gulf Coast journalists were victims themselves provided a circumstance that differed from other disasters. It's true that tragedies, such as the September 11, 2001, terrorist attacks and other natural disasters offer rare opportunities for journalists to adjust their norms to better cover large-scale breaking news events. In the case of the September 11 attacks, reporters in New York and Washington, DC no doubt felt the impact of the attacks on *their* cities, may have felt solidarity with victims, or may have personally known victims. But, by and large, they did not lose their homes, loved ones, or jobs. Nor did they have to persevere through drastic changes to their lives for years to come.

Gulf Coast journalists, however, face no quick return to the norm. They were exposed to disaster at home and at work, and that continues today. In the words of David Meeks, city editor at the *Times-Picayune*, "Journalists are always taught to keep an arms length from everything," he said. "It doesn't calculate what happens when the worst natural disaster in history of the United States hits your house, hits your family, hits all your friends, the people you love. Once that happens, no matter who you are...you will never be the same."

Journalists Are Victims Too

Meeks' comment that "you will never be the same" perhaps applies as much to his newspaper institutionally—indeed, to journalism throughout the region—as it might to him individually. This study is designed to measure whether Hurricane Katrina changed the lives and work views of Gulf Coast journalists. After all, the true signs of the journalistic impact of Katrina and the trailing Hurricane Rita must be found in the degree to which changes occurred on the media of the area.

Thus, this study is comprised of a content analysis of 667 New Orleans *Times-Picayune* and Biloxi *Sun Herald* front pages and Metro/Local Section stories. In addition, interviews were conducted with 32 editors and reporters who collectively shared the belief that better coverage of their communities post-Katrina and Rita is the tool they have to make a difference, not only for the lives of their readers, but for themselves personally.

The content study focused on types of stories being emphasized by these newspapers and on sourcing techniques used to inform those stories. To gauge the significance of this impact of Katrina on coverage in 2006, these articles were compared to coverage in 2004, the year before the storm.

Indeed, changes did occur, both in the journalists' perception of what was important and in the products they produced. One gratifying lesson learned as Gulf Coast journalists—both print and broadcast—turned their attention to the needs of their local audiences was the importance of what they were doing.

Months into post-Katrina and post-Rita life along the Gulf Coast, the *Times-Picayune* featured a story on its Front Page in January 2007 in which staff writer Coleman Warner poured out his soul to readers about how he, his wife, his teenage daughter, three cats, and his "chocolate Lab named Dutchess" had endured the perils of living in a FEMA trailer parked in front their gutted Lakeview home.[11]

Warner and his family endured a tornado that ripped through their neighborhood and threatened to turn their trailer upside down. He served as both father and part-time security guard, protecting his family as gunshots were fired in the once-secure neighborhood, now abandoned by displaced residents. And as he joked about how his family learned to shower, move, sleep, and cook meals in the confining trailer, he came to a realization. "We're part of a solidifying post-Katrina subculture, a group bound together by memories of peculiar things that come with life in a FEMA trailer, a shiny white icon of hope and loss and government bungling," he said.

Coleman concluded that "one basic truth isn't lost on us: This transitory little shelter, funky as it may seem, is our bridge to the other side of Katrina." Many reporters shared in the search for that bridge.

Reporters across the Gulf Coast recalled days without showers, changes of clothing, decent meals, or bottled water as they were scattered across their cities gathering information. In post-Katrina life, they struggled, just like their readers, viewers, or listeners, to turn their circumstances around. Spud McConnell, a radio talk-show host at WWL in New Orleans who joined the news team during the storm under a joint venture of Entercom and Clear Channel stations called The United Broadcasters of New Orleans, recalled that:

> There were so many crystallizing moments, you can't even single one out.... I think it made the callers trust us more, 'cause they knew that we were exactly in the same boat, and we were not asking anything of them, other than to listen. I lost two houses, I have all my neighbors who lost stuff, and I came from old Metairie. I didn't have 11 feet of water in my house, like the Ninth Ward. Some people down there did, but my house is gutted down to nothing. I managed to salvage some stuff, and my house is gone. I'm in exactly the same boat as everyone else.

And McConnell's loss and continued struggles, he said, gave him more credibility with the listening public: "The fact is that resonates with listeners because they know that I don't have to prove anything to them, because everything that I do to benefit me, benefits them. And I have exactly the same right as the listeners down here do to complain about our political leadership. I have every right in the world to question your decision, and your judgments if you don't have the gumption to come down here and see the hell hole we're having to pull ourselves out of."

Keeping the frustration, outrage, helplessness, or loss under control has been a challenge for these journalists whose mantle it is to help others in the news content they produce. Mark Schleifstein, a Pulitzer Prize-winning environmental reporter at the *Times-Picayune*, admitted he fights to maintain optimism so his work can make a difference in how the Gulf Coast recovers.

"We've got to be very careful moving forward," he said. "It's one thing to have the empathy because you're going through the same thing, and it's another thing there, you've got to make sure the anger you might feel because you're basically running into problems with some of those agencies yourself...doesn't immediately get thrown into your story, other than as an impetus to get to the truth of what is really going on. So that's a problem."

For Mike Hoss, evening news anchor at WWL-TV in New Orleans, his experiences and those of his colleagues have challenged the way they function as local broadcasters. He said Katrina forced them to get back to the basic roots of local journalism.

"It's made us get back to information," he said. "It's made us get back to helping people, you know, stories that have impact. It has enabled us to do so many stories with such gravity, such importance in people's lives. I have never felt more needed, as a journalist, as a reporter, in helping people rebuild their lives, in being a watchdog over government, and money and all the factors that are happening post-storm. So, it has been regeneration so to speak....It's been a rebirth and a renewing, of the basic foundation, of journalistic principles."

The Impact on Journalistic Performance

But how has Katrina impacted the way these journalists view their profession and shape the news? Normally, when journalists gather the news, facts, opinions, and information from officials and authorities are given preference.[12] The viewpoints of ordinary people become secondary in contemporary journalism or are relegated to feature and entertainment stories.[13] The personal impact Katrina had on journalists directly changed how they viewed the news and the norms of their profession.

This study indicates that, up front, it is notable that Katrina and Rita changed what these local reporters wrote about. Before the storm, the *Times-Picayune* and the *Sun Herald* were very similar to their counterparts across the nation in that stories about "cops/courts" accounted for the majority of story subjects (32 percent). But in post-storm stories, the emphasis changed with a major increase in stories classified as "other," meaning they did not belong to any of the traditional news-subject categories. Such stories were most prevalent in 2006 (at 27.5 percent), with "cops/courts" stories declining to 21.3 percent. These journalists also wrote longer, complex stories with diverse sources. Story lengths increased overall in 2006, averaging about 600 words compared to 521 words in 2004.

Another change that, not surprisingly, emerged as these journalists wrote longer stories after the storm, was an increase in the number of sources they quoted, from an average of 2.99 per story in 2004 to 4.09 in 2006. Thematic stories, defined as stories that placed isolated events in a broader context, increased in word length in 2006. Such thematic or complex stories increased in 2006 from 8 percent to 23.4 percent.

In addition, greater emphasis was placed on stories about real people, the category known generally as human interest. In 2004, 22.6 percent of the sample contained stories emphasizing human interest, a number that increased notably in 2006 to 36.9 percent. Then comes a surprise. The second most prevalent approach, the ever-present journalistic emphasis on conflict, standing at 9.2 percent in 2004, dropped to 4.5 percent in 2006.

Furthermore, a major change occurred as the journalists sought out a much wider variety of sources to contribute to the broader context, discussions of the humanity, and personal circumstances of these longer stories. Sources who previously were very limited in number—indeed, marginalized—before the storm suddenly found voices as their numbers increased significantly between 2004 and 2006.

The two newspapers found their expanded human-interest emphasis required statements from increasing numbers of individuals who were not affiliated with agencies, that is, "unaffiliated citizens." Their numbers increased from 39.9 percent in 2004 to 60.1 percent in 2006 (this was significant). In fact, individuals quoted or used in stories as the "common man/man on the street" had the most noteworthy increase from 38 percent to 63 percent (highest significance here of all sourcing).

This study also demonstrated positive relationships between the human-interest technique and these marginalized source categories. The unaffiliated citizens—persons belonging to no agency and persons identified as the common woman or man—were significantly linked to the increase in the use of the human-interest approach by these local journalists.

Interviews conducted as part of this study clearly support the conclusion that the journalists saw their jobs change as a result of the experiences they shared with each other and with those in their audiences. These took several forms.

Daily Assignments

In non-crisis times, journalists are able to plan each day based on their experience, on what their usual sources are doing and what they know of the day's schedule. They are aware of upcoming press conferences, legislative debates. Even police reporters, who must work their day around breaking news, can count on a steady stream of news releases from law enforcement on the latest crimes. But in the immediate months after the storm, even in the first few months of 2006, they could not count on a steady and predictable stream of events for the day's news.

"Normally you cover government, you cover today's event—We're going to start a new clean-up initiative, tax initiative, or the event kind of happened, a natural cycle," said Frank Donze, city government reporter at the *Times-Picayune*. "This (coverage of the storm's aftermath) is more like just finding out what is left, how much of the city is still left ... We are reacting to the stories. Every day is pretty much an adventure."

These local journalists now interweave all stories into a Katrina fabric-work, ranging from sports, to crime, to politics, to community news. Regardless of a news reporter's beat or specialty, stories are now placed in the context of Katrina. "It's to report whether things are moving or happening. That's what we've been doing, progress reports, or rather lack of progress reports," Donze added.

Converged News Beats

Since the hurricanes have linked stories together, separations or distinctions among news beats are less marked. The normal prestige that accompanies working on beats like politics now is no more than the education, health, or suburban beats. All parts of the cities' metropolitan areas have been equally affected, so the work of the journalist on each of these beats is equally respected. The journalists indicated that they share triple bylines on a regular basis. In the past, competition within newsrooms was part of the newsroom culture,[14] but Katrina increased the camaraderie in the local newsrooms with the understanding that working together is necessary to be productive and effective. They were members of a team.

"There was one police reporter with us and photographer…there was an art critic, there was a music critic, a sports editor, there were two editorial page editors, and we were just thrown into disaster coverage," said Terri Troncale, Op/Ed Page editor at the *Times-Picayune*. "So it doesn't matter what your job is on a daily basis when you're in that situation. You just have to pitch in and get things set up, and get an organized newsgathering effort going, which is what we all did, and none of us were in our regular jobs."

Gathering the News/Sourcing

One technique journalists use to shape story angles is strategically selecting and placing direct quotes or paraphrased statements. This is how they maintain distance but still get their point across. These journalists stated that since the storm, they feel less pressured into finding a source to state the obvious. They will include information as observations more often than they did before when they are unable to find a source.

"I don't know if I ask any more pointed questions, or if I am any more aggressive. I think there is just an attachment. We all are experts on this in a sense. We all went through it, and we go through it every day," Trymaine Lee, a police reporter at the *Times-Picayune* stated.

Journalists also are skeptical about reporting controversial news too quickly for fear of being inaccurate or putting their credibility in jeopardy. These local journalists believe that any new information is pertinent for their beleaguered readers, and their loyalty rests with the needs of readers first.

Gordon Russell, another city government reporter at the *Times-Picayune*, said he remained skeptical of early reports during the disaster and what he believed he saw. "We and everybody sort of over-reported the mayhem and then we did a story that debunked it," he said. "I still wonder, though, that there was more mayhem. The debunking we did was very well sourced unlike the initial stuff we did, but I'm not totally convinced that some stuff didn't totally happen....It still kind of keeps me up at night sometimes."

The Web

Print journalists often are wary of new media. They see the Internet more as a good way to hastily provide news, but one that distracts them from the skillful crafting of their writing. But, following Katrina, every one of the journalists interviewed indicated a complete change of attitude about the Internet in their daily routines. Furthermore, they embrace the web and its role in community journalism.

Perhaps more importantly, during the storm these journalists blogged, many times in first person, about the details and reports they were receiving in the earliest stages. That constant interchange continues today. Some reporters, who are not columnists, even write articles that appear in print about how they and their families are navigating the Katrina aftermath experience a year later.

David Meeks, who moved from sports editor before the storm to city editor after the storm, for example, recounts his experience with blogging: "I agree that the writing style became very personal and pointed, as (this) happened because that's just the way it was. It was so chaotic, and there were so many citizens out there trying to find people and get help, and I think the blog very much captured the mood of the people during those times. I do think our readers knew we were in a frenzied chaotic state, and the blog was almost a free-for-all for news."

Fairness, Balance, and Objectivity

While these journalists are quite clear about how their workdays have changed, they hold new contradictions about the ideals of their profession. They can agree that fairness, balance, and objectivity remain important, but they wonder how they apply in their circumstances, and they do not always agree with each other.

"I think we have gotten a little more direct...and I think it's fine," said Gordon Russell. "I still think you still have to be fair and balanced. There's no reason you have to be boring or detached. You still have to go find out the truth....One of the reasons we're boring is because we strive so hard for that objectivity that is not necessarily in service of anything or the truth. It's sort of this on the one hand this on the other hand, that and who the hell knows. I think part of our job is to go root out all of this stuff and say ok this guy screwed up, you know, and I'm going to say he screwed up, 'cause I've interviewed everybody, and he screwed up."

This is not to say journalists are not still accountable for the fairness and balance of what they put in print or on the air. And that accountability applies in the same way as has historically been the case. David Meeks stressed that the media must maintain their credibility, and one way to do this is to employ the time-tested belief that most opinion should be restricted to columns, editorials, and the op-ed page.

"In the news stories, I think we have tried very hard to be fair and objective, and while we're sympathetic to New Orleans, we want to get the facts out there," he said. "We don't want to look like the newspaper that's turning a blind eye to imperfections of our own local government or our own local effort to rebuild after the storm. You have to be fair. People will see right through it if you just write about yourselves."

But Trymaine Lee disagreed: "I think that objectivity is all very relative. Because right now, in this place and in this time, New Orleans needs the 'we' to be 'we,' the 'us' to be 'us.' Not in a sense that now you're going to take sides in these matters, but we're all deeply invested and the community needs to hear that authoritative voice that 'it's us.'"

Detachment

Can these reporters maintain the distance that normally is expected of journalists? Probably not. Within the newsrooms, reporters indicated that as time goes on, it is becoming harder to separate themselves professionally from the reality of their personal lives.

"During the immediate days, I think people had an adrenalin rush; people had their blinders on," Trymaine Lee said. "Here's the story. Everywhere you turned, you couldn't really be into this, you were right here, focused on this story. As time has gone on, we've seen people crumble emotionally. Some people have had breakdowns; some people had to medicate themselves."

Executive Editor Jim Amoss added: "[Katrina] changed the tone of what we wrote about because we knew intimately what we were talking about....We didn't have to force ourselves to be authoritative. It came naturally because it was our collective life that we led. And that has continued to this day."

Advocacy Journalism

In Amoss' eyes, his paper was forced to do the unconventional. He understands, and agrees, that a newspaper's news report always has to be fair and accurate and journalists cannot advocate in the news columns, no matter how passionately they feel. The place for advocacy is on the opinion pages of the newspapers.

"And on one occasion this past fall, we took that advocacy in the form of an editorial, and placed it on Page 1, which is normally the place of the news report," he said. "We felt so strongly about what needed to be said, at that point, mainly that the federal government owed it to New Orleans and to this region, to come to our aid in a way that they hadn't yet. And that that needed to be said as forcefully as we could possibly say it mainly on Page 1. And so some people in journalism may think of that as a blurring of the line, I think it's very clear to readers where we draw the line. I don't apologize for that."

A Personal Experience, More Personal News

The numbers and interviews in this study support the conclusion that these Gulf Coast journalists channeled the impact of the storm on their personal lives into their work. They were in fact translating their personal feelings into the shaping of stories. They would deal with their own problems at home, then go out and cover the losses of others. Their own experiences were translated into compassion for others in the emphasis they used after the storm. Said Trymaine Lee:

> You drive around the city and there is this cloud hanging over the city, and you can see how serious the situation was and is today. I think that's the part that I have to keep reminding myself of, but I have to sit back sometimes and clear some thoughts and say, this is absolutely insane. What happened was insane, the death, the destruction. Nobody asked for this to happen. You know, I think that's the part, nobody ever asked for their lives to be flipped upside down, nobody asked for the levees to break. So the people out there are still suffering even today.

The natural result was that the human-interest approach was most desirable to these journalists. They had a natural connection to the circumstances of their readers. The aftermath of Katrina and Rita is a people story, not a story about a hurricane, and journalistic sensitivity dominated the minds of these journalists who shared the burdens. This fact has had a broader impact on their work. They have expanded the human-interest approach to all types of stories, from politics to crime a year later, because the human suffering is still strong, evident in an increase in suicide and depression stories.

That's why *Times-Picayune* Executive Editor Jim Amoss will not apologize for advocating and insisted his paper will use the most effective means to do so. Most *Times-Picayune* staff members said they were increasingly convinced that their newly found passion results from reinstating themselves as the voice of the community.

"I think there was outrage involved in those headlines, and it was warranted, if anything underplayed," said city editor David

Meeks. "What happened during those first five days was absolutely ridiculous and horrendous and, based on my own reporting experience, should not have happened....At all levels there were a lot of balls dropped. And we haven't pulled our punches, but, in being balanced and careful in our criticism, have pretty much been on point."

Katrina and Rita will dominate news coverage of the *Times-Picayune* for some time to come, Amoss said. But it is clear that the hurricanes provided a new sense of purpose and meaning into Gulf Coast journalism.

"There's a lot of uncertainty in the newspaper world these days," Drew Parter,[15] photo director at the *Sun Herald* in Biloxi, Mississippi, said. "Newspapers are bought, sold, newspapers are downsizing, in some cases laying off. And a lot of journalists feel uncertain about their future. To me it really reaffirmed how important a newspaper is to the community."

Amoss recognizes the paradigm of normal journalistic short attention span. Reporters and editors come into people's lives in a moment of great crisis and write about the crisis and the trauma. Then they go home, and the next day come to the office and move on to a different assignment.

"In our case," he said, "we were in the crisis ourselves, and everything that was happening in our readers' lives was happening in our own lives individually. We were going to see people who had lost everything and who had lost their houses, and reporting about that, and then we were coming home to our spouses who had lost everything and our own houses were gone. And so I think this identification with our readers, the sharing of the same fate as our readers, changed us profoundly."

How sustained these changes will be could depend on how long Katrina/Rita is a story, or as long as these journalists embrace the lessons they learned from it. What will determine whether these changes will persist in the years to come probably goes hand-in-hand with the pace in which conditions deteriorate or improve in their communities.

Gulf Coast journalists feel compelled to adjust their former professional ideals to fit their current circumstances. Whether the changes persist or the previous norms are re-established, given the intense dedication to the community, may simply depend on what happens. How long will both residents and journalists continue to consider themselves "victims"? A return to the norm in society may—or may not—determine a full return to the norm for these journalists.

Notes

1. Staff Reports. (2006, August 9). N.O. Man Arrested After Chase. He Asked Cops to Shoot Him, Police Say, the *Times-Picayune*, Pg. B1.
2. Lang, D. (2007, December 13). Katrina Photographer McCusker Fined, Placed on Probation, PDNonline. Retrieved on March 8, 2008, from http://www.pdnonline.com/pdn/newswire/ article_display.jsp?vnu_content_id=1003685185.
3. Strupp, J. (2006). More Than $9,000 Raised For Arrested *Times-Picayune* Photographer, PDNonline. Retrieved March 8, 2008, from http://www.pdnonline.com/pdn/newswire/ article_display.jsp?vnu_content_id=1002985303.
4. Goss, A., Harwood, B., McElroy, P., & Baker, C. (2006, May 12). Profile: Katrina Photojournalist John McCusker, Voices of New Orleans, Audio Broadcast. Retrieved on September 13, 2006, from http://www.prx.org/pieces/12547.
5. As quoted in Lang, D. (2007, December 13). Katrina Photographer McCusker Fined, Placed on Probation, PDNonline. Retrieved on March 8, 2008, from http://www.pdnonline.com/pdn/newswire/ article_display.jsp?vnu_content_id=1003685185.
6. Seelye, K., Carter, B., & Elliot, S. (2005, September 12). To Publish, Not Perish: New Orleans News Media Soldier On, Ad-free, the *New York Times*, Pg. C1.
7. Mindich, D. T. (1998). *Just the facts: How objectivity came to define American journalism*. New York: New York University Press.
8. Strupp, J. (2006, August 9). *Times-Pic* Editor Comments on Arrest of Photographer, Editor & Publisher. Retrieved on September 13, 2006, from http://editorandpublisher.com.
9. Quote by Bergeron, K. (2006, October 25). Associated Press Managing Editors Conference in New Orleans video presentation, Leadership Roundtable, New Orleans, LA.
10. Quote by Varney, J. (2006, October 25). Associated Press Managing Editors Conference in New Orleans video presentation, Leadership Roundtable, New Orleans, LA.
11. Warner, C. (2007, January 17). Small Talk; Living with the Family in a FEMA Trailer Isn't a Disaster, But the Close Quarters and Lack of Privacy Won't Be Missed, the *Times-Picayune*, Pg. 1.
12. Entman, R. (1993). Framing: Toward Clarification of a Fractured Paradigm, *Journal of Communication, 43 (4)*, Pg. 51-59.
13. Entman, R., & Rojecki, A. (1993). Freezing out the Public: Elite and Media Framing of the U.S. Anti-nuclear Movement, *Political Communication, 10 (2)*, Pg. 155-173.
14. Gans, H. (1980). *Deciding what's news.* New York: Random House.
15. Quote by Parter, D. (2006, October 25). Associated Press Managing Editors Conference in New Orleans video presentation, Leadership Roundtable, New Orleans, LA.

6

Government and Journalism in Crisis:
Blame to Share

Robert Mann

In the days and weeks following Hurricane Katrina in September 2005, the national news media—the broadcast and cable networks, major newspapers, news magazines, and news services—descended on New Orleans and the Louisiana-Mississippi Gulf Coast to cover the most important news story of 2005. What they reported and showed the world was an image of a drowned city, gripped by chaos, bordering on anarchy.

As *Newsweek* later described it: "Day after day of images showed exhausted families and their crying children stepping around corpses while they begged: Where is the water? Where are the buses? They seemed helpless, powerless, at the mercy of forces far beyond their control. The lack of rapid response left people in the United States, and all over the world, wondering how an American city could look like Mogadishu or Port-au-Prince."[1]

Indeed, conditions were primitive and sometimes dangerous. Information was sketchy, especially in the hours and days after the storm hit the Gulf Coast. Perspective was often lacking. Indeed, as some journalists have reported, it was an environment strangely similar to the conditions of war reporting, with the significant absence of gunfire.

A somewhat natural consequence of this confused environment was that false rumors were rampant in the storm's immediate aftermath. Unfortunately, the media took many of these rumors and broadcast them as fact. These included inaccurate stories of rape and murder in the Superdome, citizens firing guns at their rescuers, and the toxicity of the water (some began to call it "a toxic soup").[2]

Newsweek's description was typical: "Only despair. The news could not have been more dispiriting: The reports of gunfire at medical-relief helicopters. The stories of pirates capturing rescue boats. The reports of police standing and watching looters--or joining them. The TV images of hundreds and thousands of people, mostly black and poor, trapped in the shadow of the Superdome. And most horrific: the photographs of dead people floating facedown in the sewage or sitting in wheelchairs where they died, some from lack of water."[3]

U.S. News & World Report also painted a bleak picture: "It didn't look like America, the exodus of stunned refugees wading through turbid, waist-high water, carrying only what mattered most: sick relatives, bundled babies, storm-soaked family Bibles. It looked like another country, the kind of place where armed bandits outnumber police and desperate families search garbage dumpsters for food. A place where the poorest of the poor die in the heat, their corpses ignored on the side of the road."[4]

Most of the false stories—especially those that suggested that rescuers were the targets of violence—almost certainly hindered the rescue efforts by discouraging some first responders, relief workers, bus drivers, and others from offering assistance in the crucial days after the storm. Especially appalling was the willingness of the national news media, particularly the television networks, to repeat rumors as fact. The networks, especially, appeared to embrace the philosophy that speed of reporting, not accuracy, was most important.

In the days following the storm, I and other members of the governor's staff spent considerable time contacting these networks in an attempt to correct inaccuracies. Too often, the attitude of these reporters was cavalier, as if accuracy in television news was merely a consequence of evolution or trial and error. Live television and the immediacy of the crisis, it appeared, caused some reporters to repeat rumors and hearsay as fact.

There were additional false stories that found their way into the broadcasts and onto the pages of the national media that did not relate directly to the emergency response efforts. They did, however, unfairly damage the reputations of the state and local governments. These inaccurate or misleading stories included:

- National media organizations accepted as fact inaccurate statements by Bush Administration officials who were apparently attempting to absolve themselves of blame. For example, the *New York Times*, in a September 2 story headlined "Government Saw Flood Risk but Not Levee Failure," quoted President Bush's claim, "I don't think anyone anticipated the breach of the levees." The statement was accepted as fact by the *Times*, although there was ample published evidence prior to the storm suggesting that the levees might not hold.[5]
- On September 5, *FOX News* falsely reported that President Bush had called New Orleans Mayor Ray Nagin on September 28 to plead with him to order a mandatory evacuation of the city. In fact, the massive evacuation of the city was begun a day earlier and there is no evidence supporting the claim that Bush's call to Nagin and Governor Kathleen Blanco—immediately prior to their press conference on September 28—had anything to do with the heightened atmosphere of urgency in New Orleans.[6]
- *FOX News*, on September 7, and *ABC News*, on September 11, repeated the false report that "there were 2,000 buses under water" in New Orleans that were not used for the evacuation. Reports later determined that the city owned no more than 324 school buses. If the city's public transit buses were added to that number, the total number of buses would number about 700. As it turned out, the inaccurate number appears to have originated from a *New York Times* story on September 4 that quoted Louisiana emergency planners who believed it would have required up to 2,000 buses to evacuate 100,000 elderly and disable citizens from New Orleans in the event of an approaching catastrophic hurricane. But, as the *Times* reported, this was "far more than New Orleans possessed."[7]
- *FOX News* and Knight Ridder news service reported that Governor Blanco blocked the American Red Cross from sending relief supplies to the Louisiana Superdome. Actually, the Louisiana National Guard had not prevented the Red Cross from entering New Orleans, but had advised Red Cross officials that they could not guarantee the safety of the organization's workers and volunteers should they decide to enter New Orleans. Furthermore, as National Guard officials reported at the time, the Superdome was adequately supplied with food and water for evacuees.[8]

Presenting a "Narrow and Myopic View"

Perhaps more disturbing than this incorrect information was the very narrow and myopic view of New Orleans presented by the national media. While these images were largely accurate, they did not relay the totality of the story. For that, this author, the Governor's Office, and other state officials bear some responsibility because we did not allow most reporters to enter the city and denied them access to National Guard helicopters and Louisiana Wildlife and Fisheries boats. Although hundreds of Wildlife and Fisheries boats were dispatched into the storm zone on the afternoon of Monday, August 29—just as the storm had passed over the region—no reporters, photographers, or videographers were allowed on these boats. Media were also not allowed on National Guard helicopters that began rescue missions that same afternoon.

In the days after the storm, those with automobiles were stopped at checkpoints and told they could not proceed into the city without authorization. As a consequence, the news media were generally not allowed into New Orleans from Monday, August 29 through the afternoon of Thursday, September 1. While some media organizations managed to circumvent the checkpoints, many of those who reported from New Orleans for the first three days after the storm were individuals who had entered the city prior to the Katrina and weathered the storm.

This meant that, in the days immediately following the storm, the predominant images from New Orleans were those of flooded homes with desperate residents stranded on rooftops, apparent widespread looting, dead bodies floating in flood waters, out-of-control fires, flooded school buses, the Superdome and Convention Center crowded with desperate evacuees, and confused and overwhelmed public officials seemingly unable to deal with the crisis.

Were these images accurate? They were. But they did not represent that totality of the story. For example, what is usually not reported is that, according to the Louisiana State Police, more than 90 percent of the residents of the New Orleans region were safety evacuated before the storm. Clearly, the city, the state, and the federal government did not do enough to evacuate the remaining residents. This fact is well documented and acknowledged by state and New Orleans officials.

Despite this monumental failure, the safe evacuation of more than one million citizens in less than 48 hours is remarkable. State and local officials accomplished the largest evacuation of an urban center in the nation's history and the first wholesale evacuation of a major American city since the Civil War. As the disastrous and deadly evacuation of Houston prior to Hurricane Rita demonstrated, the movement of such a large population is extremely difficult and potentially deadly.[9]

Just as significant, viewers around the country were given the impression that state officials were paralyzed by fear and indecision in the days after the storm and that no one was being evacuated or rescued, except perhaps by the U.S. Coast Guard. In truth, however, thousands of citizens were being rescued each day—in all, probably more than 50,000 citizens during the week after the storm—by the Louisiana National Guard, the Louisiana Department of Wildlife and Fisheries, and by average citizens who brought their boats into the storm zone to save lives.

That story, however, did not receive the attention it deserved because the media were not allowed on National Guard helicopters and Wildlife and Fisheries boats. The reasoning for these decisions was sound—every spot on a boat or helicopter that is occupied by a reporter or camera crew is one less person who can be saved. However, because the television audience saw very few images of these rescues, viewers got the impression that almost nothing was being done to save lives when, in fact, thousands of lives were being saved every day by the heroism of brave public servants laboring in obscurity.

The one time that Governor Blanco visited the Superdome—on Tuesday evening after the storm—she did not allow media coverage of the visit. Over the objections of her staff, Governor Blanco excluded reporters from the trip and an opportunity was missed to demonstrate the state's concern for those stranded at the Superdome. While this may seem like a small matter, it was symptomatic of the unwillingness or inability of some state and local leaders to act decisively in ways that reassured the public of their competence and compassion. The impression was left in the minds of many that government officials either did not care or were incapable of dealing with the enormous challenges presented by the disaster.

Government Caused Some of the Journalistic Problems

It is evident that the sins of the national media are not entirely their fault. In many cases, they were victims of circumstance and, in some cases, those circumstances were created by this author and his colleagues in state government. The fact that the national media often had a myopic view of the unfolding story in New Orleans had something to do with the severe constraints that were imposed upon them, including their lack of access to New Orleans and their inability to accompany state employees on rescue missions.

Another important factor contributed to the distorted view of how Louisiana state officials responded to Katrina. The evidence suggests that in the days following the storm the White House engaged in a concerted effort to shift the blame for its monumental blunders onto state and the locals. Certainly, the state deserves a great deal of the blame for its mistakes.

However, when the White House and other federal files are eventually opened to the public, I believe the record will show that while local and state officials were struggling to save lives and deal with the worst natural disaster in American history, some officials in the White House devoted their efforts not to helping state officials save lives, but rather to saving the political life of the president. They did this by shifting the blame for their failures and misconduct onto Governor Blanco and other state officials.

By Wednesday after the storm, it was clear to the governor's staff that the operation to shift blame was in full force. The governor's press office was spending the majority of its time answering inaccurate reports about the state response and responsibilities that were coming out of Washington. Time after time, the source of this misinformation was identified by these reporters as "senior White House officials."

Journalists Were Too Accepting of the White House Viewpoint

In one of the most egregious examples of this, the *Washington Post* on September 4, 2005, quoted an anonymous "senior Bush official" who alleged that Governor Blanco had not yet declared a state of emergency. The clear implication, accepted as fact by the

Washington Post (and also by *Newsweek*) and never confirmed by any source in Louisiana, was that the state of Louisiana was so thoroughly incompetent and paralyzed that its governor was not even capable of declaring a state of emergency.

"Louisiana Gov. Kathleen Babineaux Blanco seemed uncertain and sluggish," *Newsweek* reported, "hesitant to declare martial law or a state of emergency, which would have opened the door to more Pentagon help." However, as the *Post* reported on September 5 in a correction, Governor Blanco had, in fact, declared a state of emergency on August 26, three days before the storm.[10]

Some of the strategy behind the White House campaign of blame shifting was revealed by the *New York Times* on September 5, 2005. "Under the command of President Bush's two senior political advisers, the White House rolled out a plan this weekend to contain the political damage from the administration's response to Hurricane Katrina," reporters Adam Nagourney and Anne E. Kornblut wrote. These advisers, the paper reported, "directed administration officials not to respond to attacks from Democrats on the relief efforts, and sought to move the blame for the slow response to Louisiana state officials, according to Republicans familiar with the White House plan."

The reporters further observed: "In many ways, the unfolding public relations campaign reflects the style Mr. [Karl] Rove has brought to the political campaigns he has run for Mr. Bush....In a reflection of what has long been a hallmark of Mr. Rove's tough political style, the administration is also working to shift the blame away from the White House and toward officials of New Orleans and Louisiana who, as it happens, are Democrats."

As an example of administration officials who were adhering to Rove's communications philosophy, the *Times* reporters quoted Secretary Michael Chertoff of the U.S. Department of Homeland Security: "The way that emergency operations act under the law is the responsibility and the power, the authority, to order an evacuation rests with state and local officials....The federal government comes in and supports those officials." The reporters observed that this "line of argument was echoed throughout the day, in harsher language, by Republicans reflecting the White House line."[11]

In January 2007, in a speech to graduate students at the Metropolitan College of New York, former FEMA Director Mike Brown claimed that political decisions were a prime factor in the White House decision to attempt to federalize Louisiana National Guard troops during the first week of September 2005. "Unbeknownst to me," Brown said, "certain people in the White House were thinking: 'We had to federalize Louisiana because she's a white, female Democratic governor [Kathleen Blanco], and we have a chance to rub her nose in it.'"[12] As columnist Stephanie Grace, of the New Orleans *Times-Picayune* observed, "E-mails that [Blanco] long ago made public back up Brown's claim that the administration was trying to shift bad publicity to the state."[13]

When the complete story is told, I believe it will be apparent that the White House did to Governor Blanco what George Bush's operatives did to Senator John McCain in South Carolina in 2000 and to Senator John Kerry in 2004—they did what was necessary to destroy those who presented threats to Bush's political future and employed every useful tool at their disposal.

Politically, it is clear that the governor's staff, including this author, should have aggressively challenged the White House's campaign of blame shifting. The landscape is littered with the dead careers of politicians who did not respond effectively when under attack. Michael Dukakis (see Willie Horton and prison furloughs) and John Kerry (see Swiftboat Veterans for Truth) come immediately to mind. Because Governor Blanco feared offending the president at a time when the state was desperately in need of federal assistance, her communications office was forced to unilaterally disarm. While this author and others argued against this strategy, the governor was adamant that her staff not engage in what she considered partisan sniping.

Denise Bottcher, the governor's press secretary, and this author did everything possible to set the record straight. But it was impossible to persuade the governor and her husband, Raymond Blanco, to allow her staff to publicly challenge a virtual tsunami of blame that was being shifted in the state's direction by the White House, its allies, and spokespeople. It is not clear at all that the governor's staff could have competed against the resources of the White House and the federal government. The governor's communications operation had no more than five communications professionals. Even so, the unilateral disarmament in the face of a blatant, well-orchestrated effort to blame her for the Bush Administration's failures sealed Governor Blanco's fate.

State Officials Could Have Done Better

This is not to argue that the state was blameless. Clearly, state and local officials are responsible for many failures associated with the planning and response to Hurricane Katrina. Those have been well documented. In fact, because Governor Blanco is the only public official to release every document requested by the U.S. House and Senate investigating committees, more is known about the state's failures than any other government entity. This author will leave it to far more objective observers to fully analyze the state's response to Katrina.

This is primarily an effort to consider the ways that this author, as a former member of Governor Blanco's senior staff, believes that the governor and state and local officials failed in their communications with the public and the media about Katrina. The errors and misjudgments can be placed in three categories: overexposed, understaffed, and overwrought.

Overexposed

The governor's staff overexposed an exhausted governor so that the public image of her was not one of strength and resolve but rather fatigue and confusion. She should have been restricted to one briefing a day, possibly two in the early days.

Kathleen Blanco is a poor communicator. She does not speak well or with authority, except under tightly controlled conditions or in small groups. Add to that sheer exhaustion and enormous pressure, the result was a communications nightmare. The three and four press briefings a day during the week after the storm—while not problematic for a strong communicator—exposed her to questions she was often not prepared to answer and highlighted her weaknesses. As a consequence, the strain and confusion was all too apparent in her voice and her demeanor. While the demands of the local and national news media for access to the governor were enormous, this author accepts responsibility for succumbing to that pressure in ways that did not serve her well.

Understaffed

The state clearly had its public affairs resources in the wrong place. With few exceptions, all of Louisiana's public information officers were in Baton Rouge, which was the site of a large part of the story. But a much more important story was unfolding 80 miles to the south and the state had almost no one to help manage communications for the governor, the National Guard, the state police, and the Department of Wildlife and Fisheries. As it turns out, even had public affairs officers been sent into the storm zone, they would have been isolated and out of touch much of the time because of the total breakdown in communication between New Orleans and Baton Rouge.

Furthermore, as already discussed, the state *prevented* access to New Orleans when it should have *controlled* access to New Orleans. But since the state government lacked the resources and the communications to control that access, it was prevented for the first three and a half days after the storm—and, therefore, much of the story of the state's role in the rescue was not reported, or at least was not recorded by still and video photographers.

It was clearly wrong to not make accommodation for media on boats and helicopters. Saving lives was clearly the priority; but public confidence in government also is crucial. During times of crisis and danger, the public deserves to know that public servants are working effectively on their behalf—and *showing* them is far more effective than merely *telling* them. In the months following Katrina, state officials began discussing ways to ensure that media representatives can be accommodated on boats and helicopters in the event of another hurricane. Should another strong hurricane strike south Louisiana, the media will be allowed to accompany rescuers on their missions.

Overwrought

In the confusion and stress of the aftermath, the image shown to the nation and the state was that of a government completely unable to deal with the crisis. Much of it was, of course, the very valid criticism of the city and the state's inability to evacuate those without transportation and who were later stranded in the dome and the Convention Center. A substantial part of it was the very effective blame game played by the White House at the state's expense.

However, the negative image of the state was fatally compounded by the inability of the state, the federal government, and the city of New Orleans to work together in the days after the storm. For that, New Orleans Mayor Ray Nagin was largely responsible. At the time, members of the governor's staff were persuaded that he was working in concert with the White House in its effort to smear the state.

The record will show that Governor Blanco never publicly uttered an unkind word about Mayor Nagin in the weeks after the storm. When he was having what many believed was a mental breakdown, the governor resisted the urge to feud with him because she believed it would only further undermine public confidence in the response and recovery. The fact is, however, that the inability of the governor and the mayor to communicate in the days after the storm reinforced an image that persists to this day—and that is one of failure and overwhelming incompetence.

Because federal officials—particularly officials working for the Federal Emergency Management Agency and the Department of Homeland Security—were actively engaged in the White House smear campaign, the relationship between the governor's office and federal officials was severely strained and added to the dysfunction of the coordinated response.

The damage to Louisiana's public image is insignificant compared to the enormous suffering and loss endured by hundreds of thousands of Louisiana citizens. This analysis is not intended to diminish or detract from that suffering. It is, instead, an attempt to explain some of the ways that this author and other communications professionals could have better discharged their responsibilities and helped the local and national media provide a more enlightened view of Louisiana and its response to Katrina. It is presented in the hope that other public officials and communications professionals will learn from our mistakes and not repeat them should another disaster befall this state.

Most disasters do not hold the media's attention for long. There is always another disaster or scandal to divert the media and the public's gaze. In this instance, and to the national media's credit, many major news organizations have remained interested in the recovery. National news organizations—including NBC News, CNN, and the *New York Times*—have demonstrated their commitment to the story by establishing bureaus in New Orleans. These and other organizations have dedicated themselves to doing what they can to keep the public informed about the state of the recovery.

In some small way, this continuing commitment to covering this important story serves to atone for some of the media's errors in the early days after the storm. While the mistakes of those early days cannot be undone, it is possible to correct the record by giving the nation the broader, more comprehensive view of New Orleans and the Gulf Coast that was woefully missing in September of 2005.

Notes

1.   Thomas, M. (2005, September 12). The Lost City, *Newsweek*, Pg. 42.
2.   For a sampling of "toxic soup" references, see: Knickerbocker, B., & Jonsson, P. (2005, September 8). New Orleans toxic tide, Christian Science Monitor, Pg. A1; Kushner, A. B., (2005, September 12). After the Flood, *The New Republic*, Pg. 42; Coy, P., Foust, D., Woellert, L., & Palmeri, C. (2005, September 12). Katrina's Wake, *Business Week*, Pg. 32.
3.   Thomas, M. (2005, September 12). The Lost City, *Newsweek*, Pg. 42.
4.   Mulrine, A., Marek, A. C., Brush, S. (2005, September 12). To the Rescue, *U.S. News & World Report*, Pg. 20-26.
5.   Shane, S., & Lipton, E. (2005, September 2). Government Saw Flood Risk but Not Levee Failure, *New York Times*, Pg. A1; Lipton, E. (2006, February 10). White House Knew of Levee's Failure on Night of Storm, *New York Times*, Pg. A1.
6.   Hume, B. (2005, September 5). Who's to Blame and Who Isn't to Blame. Special Report with Brit Hume, Fox News. Retrieved on April 13, 2008, from http://www. foxnews.com/story/ 0,2933,168582,00.html.
7.   Hannity, S., & Colmes, A., (2007, September 7). Should New Orleans Be Rebuilt? Hannity & Colmes, Fox News transcripts; Stephanopoulos, G. (2005, September 11). Relief Effort: New Orleans Update, This Week with George Stephanopoulos, ABC News Transcript.
8.   O'Reilly, B. (2005, September 6). *The O'Reilly Factor*, Fox News Transcripts.; Hume, B. (2005, September 11). Fox News Sunday, Fox News Transcripts.; Media Matters (2005, September 11). Knight Ridder, Palm Beach Daily News, Fox News' Hume Repeated Misleading Red Cross Story, Shifted Blame to Blanco, Palm Beach Daily News. Retrieved on April 13, 2008, from http://mediamatters. org/items/200509120001.
9.   Harden, B., & Moreno, S. (2005, September 23). Thousands Fleeing Rita Jam Roads From Coast, the *Washington Post*, Pg. A1.

10.   Roig-Franzia, M., & Hsu, S. (2005, September 4). Many Evacuated, but Thousands Still Waiting, the *Washington Post*, Pg. A1.; Roig-Franzia, M., & Hsu, S. (2005, September 5). Corrections, the *Washington Post*, Pg. A2.; Thomas, M. (2005, September 12). The Lost City.
11.   Nagourney, A., & Kornblut, A.E. (2005, September 5). White House Enacts a Plan to Ease Political Damage, *New York Times*, Pg. 14.
12.   Toosi, N. (2007, January 20). Former FEMA head Brown says party politics played role in Katrina response, Associated Press Worldstream.
13.   Grace, S. (2007, January 23). Brown's troubling remarks deserve airing, *Times-Picayune*, Pg. 5.

7

Journalism Defines the Issue:
Coastal Erosion

Jane Dailey and Lisa K. Lundy

It took more than 7,000 years for the Mississippi River to build southern Louisiana. It took humankind 70 years to undermine the construction process. And it took hurricanes Katrina and Rita to show the world just how fragile the coastal ecology can be.

That is the focus of a series of stories appearing in the New *Orleans Times-Picayune* in 2007. The series concluded that the sea would reclaim New Orleans unless Louisiana found a solution to coastal erosion within the next ten years.

Most coastal erosion stories were not focused this way prior to Katrina and Rita. If journalists thought about coastal erosion at all, it was most commonly considered to be someone else's problem. And most people didn't think about it all because it was a subject of little attention in both print and broadcasting.

The lack of attention to the issue was so obvious that the state of Louisiana created a public relations campaign in 2002 in the hope of bolstering national support—and, in the process, federal construction dollars—to rebuilding the rapidly disappearing marshlands. The campaign, called "America's Wetland," attempted to portray coastal erosion as not just a problem for coastal states but as a problem affecting the nation.

This study began as an attempt to determine the effectiveness of that campaign. Arrival of the hurricanes, however, provided a chance to extend the study and to examine how they impacted national awareness of coastal ecology. At the heart of this analysis was whether messages about coastal erosion were likely to create a national sense of responsibility or whether they would continue to be perceived as another person's problem.

The study found newspaper coverage of coastal erosion was scant but increased rapidly following the hurricanes. It also found that Katrina and Rita did not change the way the media focused on coastal erosion—the focus remained on coastal erosion as a local issue rather than as a national problem. It also found that government sources were most frequently cited by media and that environmental groups played only a negligible role in coverage and, thus, by implication, were ineffective in setting the agenda.

The "America's Wetland" campaign focused on the economic role that coastal marshes, such as in Louisiana, and coastal activities, such as offshore drilling, had on the nation. It pointed out that marshland played a critical role in the lifecycle of the nation's seafood industry. It noted the contribution of the offshore oil and gas industry to keeping the nation's homes warm and cars fueled. And it emphasized the protection the coastline gave to major metropolitan cities, such as New Orleans, and focused on the disruption, regionally and nationally, that would occur if a hurricane were to destroy New Orleans or damage the nation's most important transportation highway—the Mississippi River.

The campaign had limited success in selling the story as a national issue. In-state coverage exceeded out-of-state coverage by 247 to 94 stories during the years of this study, from October 2002 when the campaign began to August 2005 when Hurricane Katrina attacked the Gulf Coast. The study included five national newspapers, six regional newspapers, and gave principal focus to two Louisiana newspapers.[1]

Admittedly, coastal erosion stories are a hard sell. Journalists favor simple stories that have a starting and an ending point and an immediate impact on readers. Coastal erosion is a highly complex story, the starting and ending points are far in the past and future, and the impact on readers may not come for generations upon generations. Additionally, few reporters have the scientific background to understand the issues, and few of those in the audiences want that depth of knowledge.

Further complicating the matter is that the science behind coastal ecology is theory-based, and the theories often are in direct conflict. About the only news value coastal erosion has is that the political, economic, and social costs are enormous, but even that is a problem. The cost of fixing the problem is beyond the comprehension of most, and few people are willing to take ownership of what they believe is someone else's problem.

Katrina and Rita, however, had fabulous news value—widespread destruction, human despair, political controversy, and the failure of governmental institutions at every level, all wrapped up in a comprehensible timeline. No American city had ever been destroyed to the extent of New Orleans. Yes, the Great Earthquake of 1906 pretty much leveled San Francisco and probably caused much greater loss of life. And true, the plagues and subsequent Great Fire in 1665 and 1666 destroyed 20 percent of London's population. But no sound bites or video highlighted those events, and they were hardly contemporaneous. Only 9/11 compared in terms of impact, number of deaths, and citizens affected.

Perhaps the greatest impact of Katrina and Rita was in what may be called the disaster quotient. Stories prior to the hurricanes tended to shy away from end-of-the-world type statements, whereas stories after Katrina and Rita tended to focus more on the potential for even greater disasters. This held true for both those stories that focused on local impact and those that emphasized national impact. The stories that concentrated on national impact, however, tended to view coastal erosion as part of a larger picture of global warming.

Coverage Focused on Suffering, Loss, and Disaster

The media depiction of local impact epitomized three main sub-themes: local conflict, human suffering or personal loss, and a sense of widespread disaster. The sense of disaster dominated all other sub-themes. This is likely because of the effects of two years of active hurricane seasons that affected several states along the United States Gulf Coast. Katrina, after all, was followed by

Hurricanes Rita and Wilma in 2005. Several of the depictions of frames of local impact in stories about coastal erosion follow:

> Today, if a Category 1 hurricane were to hit Galveston Island, all but the western-most part would remain above water. If Galveston sinks as much as it did in the last century, by 2100 nearly the entire island would be inundated. (*Houston Chronicle*, Dec. 25, 2005)

> Experts have warned about New Orleans vulnerability for years, chiefly because Louisiana has lost more than a million acres of coastal wetlands in the past seven decades. (*St. Petersburg Times*, Aug. 29, 2005)

> Historically, the Louisiana coast enjoyed a measure of protection from hurricanes. During the past century, coastal Louisiana has lost much of [its] natural storm shield. (*USA Today*, Oct. 3, 2005)

> The refusal to build spillways and reservoirs exacerbated the effects of the 1927 flood just as coastal erosion and the blazing of shipping channels presumably contributed to Katrina's destruction. (*Los Angeles Times*, Aug. 31, 2005)

> He [Donald Bourg] has watched the earth—hundreds of square miles of stunningly rich Louisiana land—disappear before his eyes. The territory around Bourg's tiny hamlet of Dulac...is in the heart of a vast disappearing act. (*Washington Post*, July 13, 2003)

> Tropical storms and hurricanes change the vegetation line so that a house that was once behind the line of vegetation ends up on the public beach...The problem is exacerbated by the fact that the Texas coast is eroding at a rate of about 15 feet a year. (*Houston Chronicle*, June 8, 2004)

> This surfer's paradise is at the center of a philosophical tug-of-war ...between those who put up the seawall to save the road and public access, and those who believe seawalls are an abomination, that nature should be allowed to have its way. (*San Francisco Chronicle*, Dec. 26, 2003)

> The campaign [America's Wetland]...was supported by a host of major corporations, particularly large petrochemical and energy companies—entities that have been accused of contributing to the destruction of Louisiana's coastline through dredging and ship traffic. (*Houston Chronicle*, Aug. 15, 2004)

National Stories Emphasize Global Warming

The media depiction of national impact was far outnumbered by articles discussing local impact. That is not surprising. Journalists tend to focus on issues with local impact, and this creates an enormous challenge when trying to cast a local issue as a national problem.

The depiction of national impact could be seen as two main sub-themes: global warming and the extent to which large numbers of people could be affected. Of the articles discussing national importance, most were related to global warming and within them discussed coastal erosion problems. Several of the depictions of frames of national impact in stories about coastal erosion follow:

> Nearly a third of the fish harvested by weight in the lower 48 states comes from Louisiana. The state's commercial seafood harvest is about $343 million annually, and recreational fishing is a $944 million industry. (*Houston Chronicle*, Oct. 4, 2003)

> We will renew and restore New Orleans and the region because its existence is dictated by the needs of U.S. Commerce. The question is not whether Americans can afford to raise up Louisiana's economy; it is whether American can afford not to. (*Washington Post*, Oct. 1, 2005)

> With an issue so important to the future of Alaska, the United States and the world, continued "paralysis by analysis" is unforgivable. (*USA Today*, July 19, 2002)

Katrina and Rita had an impact beyond the emotional content of the language, however. The impact also could be measured in terms of number of stories that focused on coastal erosion. Twenty eight percent of all articles examined were published after the hurricanes compared to 14 percent the previous year, doubling the amount of the previous year's coverage in just six months following the hurricanes. Perhaps of more significance, however, was the prominence given articles on coastal ecology. The majority of stories written before Katrina appeared on the inside sections of the newspapers. After Katrina, more stories appeared on the front page and slightly more on section fronts.

Government sources were most frequently used in stories about coastal erosion, followed by academic scientists and researchers. Of all the articles, spokespersons from "America's Wetland" campaign were the least used source even though 40 percent of all articles were about Louisiana coastal erosion. In addition, only 10 percent of the articles mentioned or cited environmental advocacy groups such as the Sierra Club, National Defense Fund, or Nature Conservancy.

Coastal Erosion Still Considered a Local Issue

It is clear from the data that Louisiana's attempts to make coastal erosion a national economic issue was less than a resounding success. The majority of articles continued to stress the local impact of coastal erosion. Overall, 32 percent of all articles published before Katrina and Rita focused on local importance or local impact of

Figure 7.1
Frequency of Coastal Erosion Stories Pre- and Post-Katrina

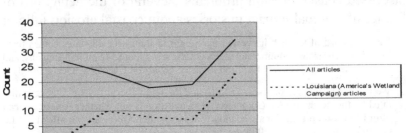

coastal erosion. This compared to only 10 percent that framed it as a nationally important issue. In fact, politics was significantly more dominant in pre-Katrina/Rita coverage with as much as 25 percent of all stories focused on politics.

After the hurricanes, the number of articles focusing on local importance continued to increase to as high as 42 percent, while stories focusing on national importance dropped to only 3.7 percent, even after the nation's attention was redirected toward Louisiana and other vulnerable coastal areas. Overall, local importance both before and after the hurricanes, including those stories specific to Louisiana and the "America's Wetland" campaign, consistently out-framed other topics including environmental impact, which was reflected in less than 3.3 percent of all articles.

Even those stories mentioning the "America's Wetland" campaign followed that pattern, although slight improvement was noted in the number of stories focused on economic importance, responsibility, and national importance. Environmental impact was the least used focus.

The dominant focus after Katrina continued to be local impact (Figure 7.2). The only major change in the focus used in news stories pre- and post-Katrina was that stories about Louisiana's coastal erosion more frequently cited governmental failures after the hurricanes. This may be because of the amount of finger-pointing that occurred among political leaders following Hurricane

Figure 7.2
Focus of Coastal Erosion Stories Pre- and Post-Katrina

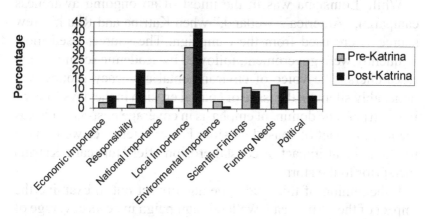

Katrina and a more intense search for answers about why and who allowed this disaster to occur.

The "America's Wetland" campaign, by its very name, was centered on raising national awareness and gaining federal support for Louisiana's coastal erosion problem. However, the data show news coverage of Louisiana's campaign rarely depicted impacts to the rest of the nation despite the title and the attempt to sell the problem as a national issue.

The majority of stories stopped short of describing why the nation should assume ownership. Only a few newspapers mentioned the campaign. And even those that mentioned the campaign focused primarily on establishment of the campaign itself or the issues that impacted Louisiana, not on the national impact. This pattern, established prior to Katrina and Rita, held true after the hurricane as well, as Figure 7.2 notes. In fact, post-Katrina stories were even more focused on local impact than pre-hurricane stories, and national impact was even more diminished.

Louisiana has been publicizing its coastal erosion problem for years, trying to establish the issue on the national news agenda. The amount of national coverage of coastal erosion increased following hurricanes Katrina and Rita. The prominence of coverage of coastal erosion did change, moving from inside the paper to front and section front pages.

Environmental Voices Silent in Coverage

While Louisiana was in the midst of an ongoing awareness campaign, "America's Wetland," when Katrina and Rita hit, few sources were used from the campaign. The sources used most frequently were government, followed by academic scientists and researchers. The voice of environmental advocacy groups was noticeably silent on the issue of coastal erosion in the days following Katrina. The dominant emphasis in coverage post-Katrina was the local impact of coastal erosion. Fewer references were made to the national impact of coastal erosion after Hurricane Katrina than prior to the storm.

In beginning of this study, the authors set out to examine the impact of the "America's Wetland" campaign in news coverage of coastal erosion in Louisiana. The authors reluctantly concurred that while the "America's Wetland" campaign may elevate the perceived significance of coastal erosion in Louisiana, it was likely that only a major hurricane might really draw the nation's attention to this environmental issue.

As such, it was stunning that the coverage of coastal erosion in Louisiana post-Katrina/Rita failed to portray coastal erosion as an environmental issue or an issue of national importance. While coastal erosion is certainly an important issue for Louisiana and coastal communities in other states, important national impacts of this issue involve the nation's economy, various seaports, the seafood industries, and the environment. While it is difficult to communicate how coastal erosion on the Gulf Coast affects large numbers of people throughout the country, such communication is needed to facilitate the sense of national responsibility needed to prompt policy change and federal help.

Note

1.   Newspapers included in this study were: national newspapers, the *New York Times*, *Washington Post*, *Los Angeles Times*, *Christian Science Monitor*, *USA Today*; regional newspapers, *Chicago Sun Times*, *Houston Chronicle*, *Boston Globe*, *San Francisco Chronicle*, *St. Petersburg Times*, *Atlanta Constitution*. A Lexis Nexus search was conducted of all Louisiana newspapers, but focus was placed on the *Advocate* of Baton Rouge and the *Times-Picayune* of New Orleans because they were the sources of most of the coverage.

8

Public Relations: In the Eye of the Storm

Lisa K. Lundy and Jinx C. Broussard

The relationships between journalists and public relations practitioners were never clearer, or more chaotic, than during and in the aftermath of hurricanes Katrina and Rita. PR practitioners were functioning under the same unbelievable circumstances as their media colleagues. They and their families were participants in the story. Many evacuated. Some remained as their families rushed to other parts of the country. Others had evacuees living in their homes. They had no electricity, few means of communication, their streets were inundated, and often their sources of information were unavailable, unreachable, or simply knew nothing.

Yet, they still had a job to do. They were responsible for communicating timely and accurate information that their publics needed to make prudent decisions and for simultaneously highlighting the efforts of their organizations in preparing for the storm, during the storm, and during the recovery. Fellow employees—often in physical danger—needed to know what was going on within the company or the agency. Journalists clamored for information that was simply not available. Often, they had no plan at the ready for dealing with such an emergency as that of Katrina and Rita.

As Francis Marra said in 1998: "Crises, in almost all circumstances, immediately trigger a deluge of questions from an organization's many different publics. Reporters, employees, stockholders, government officials, and local residents all want—need—to know: What happened? Who did it happen to? When? Where? How? Why? Organizations that wait to answer these questions often suffer unnecessary financial, emotional, and perceptual damage. The ability to communicate quickly and effectively is clearly an important component of successful and effective crisis management."[1]

This study investigates those roles, efforts, and experiences of these PR professionals. Using the contingency theory of public relations, the researchers sought to understand the extent to which public relations professionals balanced accommodating their publics and advocating on behalf of their organizations in the days following the storms. The researchers especially were interested in how public relations professionals, during those tumultuous times, balanced their needs to accommodate their publics and advocate on behalf of their organization.

Specifically, the study asked to what extent the public relations practitioners enacted crisis communication plans prior to and after the storm, what their experiences and challenges were, and what strategies they employed in interacting with various publics in the days following the storms.

In search of this information, from October 2005 to January 2006, the researchers conducted in-depth interviews with fourteen public relations professionals whose organizations were directly impacted in Baton Rouge and New Orleans, Louisiana.[2] Some of the interviewees asked that they not be identified.

The Situation

In the face of hurricanes Katrina and Rita, the professionals learned quickly. Most worked around the clock for the first few days, and some kept long hours for weeks and months after the storm. Bob Johanessen, communications director for the Louisiana Department of Health and Hospitals, worked 20-hour days for the entire month of September. Another professional had a similar experience.

They had no choice. Even as they faced their own personal emergencies, business had to continue. They had to handle the emergency communications all day every day, one-on-one interviews, fact checking, and at times media briefings every two hours.

The challenge, Kendall Hebert, director of public relations for the Louisiana Capital Area Chapter of the American Red Cross, said was juggling a massive number of phone calls and trying to give out accurate information with an upbeat and helpful attitude.

And this was being accomplished while communication was a nightmare. The systems they usually counted on did not work and people they needed had evacuated. Some managed to function in spite of the lack of a coordinated communication plan, while others simply evaluated and their operations shut down for as much as three weeks.

Dealing with the Media

One thing did not change—in fact was intensified—in the wake of Katrina and Rita. Journalists became a primary public for the PR professionals who often were pressed to change their normal focus in a city with limited electricity, media being published from other locations, impossible transportation problems, and the scattering of both their own staffs and the journalists themselves.

As a result, participants were pressed to use radio and the Internet to reach publics cut off from traditional media outlets by geographic distance, lack of power/cable, or interruptions in mail service. They were overwhelmed by requests from journalists facing the same problems. They quickly learned, however, that those journalists who remained functional must be their primary public because the news organizations from around the world were the main conduit through which they could communicate with evacuees, citizens and constituents who remained in the cities, tourists in New Orleans and those planning to travel to the city, donors, volunteers, the federal government, the nation, and the world community

Among their first needs after the storms, of course, was to determine what information the media were seeking and how best to deliver that information. "We monitored the news like crazy," said Larry Lovell, an account supervisor for Peter Mayer Public Relations who represented the New Orleans Tourism Marketing Corporation. Participants said the media were particularly interested in how many lives were lost in the storm. They wanted to be embedded in rescue operations to gather footage and conduct interviews. They wanted to interview people who had been evacuated. They wanted to tour evacuee shelters and treatment facilities. They wanted tales of heroic efforts and hours/availability of organizations for customers.

One professional, director of communications and public relations for an educational institution, remembered her problems in dealing with journalists who requested to visit her campus to take pictures when it was and remains a secure construction site.

These journalistic needs, as legitimate as they were, posed several problems for the PR professionals. They were challenged to balance the media need for information with the evacuees' right to privacy. They were trying to accommodate multiple publics and balance the needs of each. And they had constant problems of tracking down information themselves. At times, the priorities of saving human lives or dealing with people who were in physical danger interfered with their abilities to answer journalists' questions.

A major challenge, the professionals said, was the media's focus on the number of casualties. Journalists wanted numbers, and especially in the days immediately following the storms, such information was not available.

"We kept telling the media, 'We are rescuing live people. I can tell you how many real people we've picked up off of rooftops. I can't tell you how many bodies have been found because they haven't started counting,'" one said. "And they refused to accept that. It was just anarchy."

But the problems with which the PR professionals had to deal on an hourly basis were more than inconvenient, inappropriate, or impossible questions. As was the case with everyone, they were forced to cope with misinformation and rumors from the public, from their own colleagues, and certainly from journalists seeking to either confirm or elaborate something picked up on the street.

"The main problem we had during the chaos is people would call in saying, 'I heard this,'" Hebert said. "And I informed the media quickly: Do not take the public's word as truth because it's not happening. Unless you hear it from me or public affairs from the operation, don't announce it. I'll come on and tell it to you live if you want me to."

Likewise, what they considered to be imbalance and sometimes downright inaccuracy occurred in the media as journalists sought out individuals to speak for publication or on air.

Lovell was disturbed about what he considered to be inadequate, perhaps even biased, coverage by some in the national media. He thought it to be poor journalism that he found frustrating when media organizations did not balance coverage of self-described "experts" with coverage of credentialed experts. "If somebody makes an inaccurate statement, and we've seen stories with a proven inaccurate statements," he said, "you see stories that are perpetuated in a negative slant that are based on this inaccurate information."

"Local Media Get It"

For many of the participants, working with the local media was less of a challenge than working with the national and international media. They found it a challenge to meet the needs of national correspondents who did not have local backgrounds as well as answering requests from international news outlets, including reporters from Japan, Germany, Australia, China, Israel, Denmark, and France.

The ease of working with local media appeared to be the result of working relationships built over time. Local media also were useful to participants for responding to rumors and misinformation. As put by Lovell, "Local media get it."

Johanessen described his reliance on local media to disseminate information quickly. "We just stood in line outside the door of Channel 9 and talked live.…There were no time constraints or anything. Then we'd just send them over to Channel 2 and do the same thing."

Principal among local media for reaching their constituencies, the PR professionals said, was radio. Radio stations—the media being used by citizens without power, Internet access, or cell phones—quickly became the most effective way to reach publics. Chris Spencer, senior communications manager for JPMorgan

Chase, said his next crisis plan "will factor in radio" because in an emergency (such as Katrina), "the media of choice...without question, is radio." This is in contrast to normal practice because it is not normal for PR operations to use radio.

"Going on live radio, all the warning bells go off big time. So we don't do a whole lot of live radio; chances are, someone on hold on the other line is mad about their checking account," he said.

In addition, the PR professionals found value, when they were available, in using newer forms of communication. They used the Internet and e-mails to provide information to internal publics and others who sought timely and accurate information but were not in the range of the broadcast or local and state print media.

In short, the disastrous, chaotic, and limiting setting imposed by hurricanes Katrina and Rita, during the actual days of the storms and for weeks later, forced PR professionals, often working without preconceived planning, to be innovative and creative and to take advantage of any opportunity provided. Thus, they were both proactive in disseminating information and reactive in responding to media inquiries. They used any available medium and worked in any setting that was momentarily appropriate.

In general, participants were stretched to communicate creatively, as they were unable to rely on established channels of delivering information via e-mail, fax, or phone. After unsuccessfully trying to call TV and radio stations with updates on his organization, Spencer said, "The most effective way for me to deliver news was to get in my car and drive around the city." In this way, he was able to help bankers inform customers that Chase branches were open in nearby cities for longer hours, and even on Sunday, and that the bank would honor customers from other banks. Marketing also created printable flyers for branches to post on their doors. That information was vital to displaced citizens who needed access to funds.

As participants juggled media requests, they also maintained and, in many cases, increased communication to their employees and other internal constituencies. Several were communicating to displaced employees and offering important assistance.

For one public relations professional at a New Orleans university, students were dispersed all over the country, and the school's database was destroyed in the flooding. The university set up a web site through which students could obtain information about the school's policy for accepting course credits from the more than 200 universities that had agreed to host them. Students, faculty, and staff also could register on a contact form, providing address, phone numbers, and other relevant information. Later, employees were notified about whether they would remain on the university's payroll.

Accommodation or Advocacy?

Public relations professionals had their hands full as they sought to address the aftermath of hurricanes Katrina and Rita, a crisis situation that threatened the viability and reputation of local and state government and related agencies, businesses, and nonprofits.

Based on the interviews, it is clear that a focus of the participants in the weeks following Katrina and Rita was accommodating their publics. Findings indicate that whether they functioned as managers or technicians, the public relations professionals primarily sought to accommodate their publics by utilizing television, newspapers, radio, the Internet, and even cell phones (via text messaging). The public relations efforts also involved communicating what the organization was doing on behalf of its publics—advocacy, to a lesser extent.

As participants reflected on what they did to accommodate their publics, it seemed natural to highlight their efforts in future promotional communications. Thinking about this advocacy, however, was more natural to some than to others. One public relations manager stated that his organization took footage of its employees working to serve customers during Katrina and planned to use the footage later in commercials.

"It's not anything we have pushed, but how do we take advantage of that?" Spencer said. "Should we take advantage of that? Up to this point we really haven't. I prefer to let reporters draw their own conclusions. A lot of the things we could do, no one else could do, so we showed what we could do. Does it become part of an ad campaign? It might. There's been a lot of talk in the corporation about it, but not on our end. We think we've done enough."

It has been said that public relations is performing and then communicating the substance of that performance. This seems to be particularly true in a crisis. What happened was a convergence of accommodation and advocacy in the wake of the storm. In other words, the best way to advocate for your organization *is* to accommodate your publics. For example, in dealing with the media, participants spent a great deal of time with reporters and writers, helping to find the information they needed.

Most frequently, the professionals considered themselves information sources. They devoted time to providing facts to journalists who requested information. Often, they would make special efforts to address the needs of citizens through the media, both by initiating contacts and by responding to such questions as to where people could put their children in school or what to do with their light bills.

Hebert mentioned incorporating advocacy into media relations by expressing the needs of the organization. "Even though we're working a disaster, people are affected. That's when the public sees what we are really about..." Hebert said. "In every interview, you turn your interview to say, 'We still need help.'...We go into the shelter and show, 'This cot is because somebody donated $200.' 'This pillow is from a donation.'"

Many organizations scheduled town hall meetings in various cities to reach displaced publics. Those meetings not only provided updates on the process and progress of recovery and were avenues to obtain feedback for publics, but also enabled the organizations to inform their publics of how they were working for them. Some professionals wrote op-ed pieces to highlight their organizations' efforts on behalf of their publics.

In the immediate aftermath of the storm Lovell and his organization, recognizing its crucial role in the recovery, pulled together its partner publics, including state and city officials, to fashion a plan to guide their effort. The organization created a media center and held press conferences with government officials and partners to provide information that other publics needed in order to plan and also highlighted their proactive stance.

Developing a Good Plan

One point that became immediately clear during the interviews was that the public relations professionals in this study were no different from their journalistic counterparts in Louisiana and around the country. Even though everyone had talked about the certainty that "the big one" was going to hit New Orleans some day, they had no plan, or at least no plan that was capable of dealing with such magnitude and complexity. Their pre-Katrina and pre-Rita crisis planning was either minimal or simply not very helpful.

And, yes, hindsight is 20/20. The public relations professionals, in the aftermath of the storms, now affirm to prepare for the next story, to be ready, to make clear who is responsible for what, to have the assignments made, and to know whom to contact. Even more specifically, it must be clear who is to be the organization's first responder and what that person needs to do to be immediately ready.

When discussing crisis planning and its relevance in the storms' aftermath, the practitioners indicated that relationship building with various publics had been an integral part of crisis planning. Of course, no plan can anticipate every contingency. But careful considerations of their experiences with hurricanes Katrina and Rita made it clear to the PR professionals in this study that the effectiveness of their efforts during a crisis depended on several essential factors, such as:

- Effective public relations personnel should be either part of the decision-making team of their organization or at least work closely with the decision-making process. This is the only way they will be able to advise decision makers on courses of action and how to communicate with diverse publics.
- An effective plan must assure that it prepares for all magnitude of crises, including a once-in-a-century storm.
- The plan should address what-if scenarios and such basics as where the staff will get food and other necessities or where their relatives may evacuate. This is a major factor in addressing the needs of the employees who remain during a crisis.
- The plan should not only be in an office file or on an office shelf, but should be portable and easily accessible via backup on a computer disk or other mobile technology, on an alternate server, and in a binder that evacuating practitioners may take with them.

- Information in the plan should include the names, telephone numbers, email addresses, and other contact data for decision makers, staff, and key government, community, and business leaders. These key individuals also should provide the public relations practitioners the destination where they plan to evacuate and the names of relatives or friends at that location.
- The crisis communication plan should mandate that all public relations practitioners be clearly designated as *essential* personnel, and it should clarify responsibilities of each staff member.
- An effective plan should indicate that essential personnel are to remain at their assignments and, in the event evacuation is necessary, should designate destinations where staff may establish emergency operations centers through which to enact coordinated communication and other crisis management activities.

Even in the absence of such a plan, many public relations practitioners, as did their journalistic counterparts, met at least some of their responsibilities. Several professionals indicated they communicated with employees and provided information customers needed to prepare for the storms' onslaught and aftermath. But to the degree that they were able to do anything, many acted not on a preconceived plan but on the professional instincts developed over years in the business.

Even in the face of their own suffering and inconvenience, they automatically did what they had to do, what their experience indicated was needed, and what they had done in other situations. In the face of the kind of public disaster wrought by hurricanes Katrina and Rita, they sought to emphasize the "public" part of public relations.

Notes

1.   Marra, F. (1998). Crisis communication plans poor predictors of excellent public relations. *Public Relations Review, 24 (4)*, Pg. 461.
2.   Interviews were conducted with seven females and seven males representing varying levels of state and local government, quasi-governmental agencies, non-profit organizations, healthcare and educational institutions, and a variety of service organizations and corporations. Professional experience in public relations ranged from five to thirty years, but few had dealt with PR in a major crisis prior to hurricanes Katrina and Rita.

9

Public Polls: Journalists Get Good Marks

Kirsten Mogensen and Ralph Izard

Few people, including journalists themselves, are likely to argue that the media covered the terrorist attacks of September 11 and the hurricanes of 2005 with perfection. Academic studies and subsequent analyses—indeed, work by the very journalists involved—demonstrate that the coverage at times was incomplete or off-target, sometimes factually inaccurate, often overly sensational, and perhaps even insensitive.

But in the aftermath, a beleaguered public seems not to have gotten these messages of fault. Say what you want about public opinion polls, but they present a consistent finding in general of public approval of journalists' work during those very difficult times. And it's not just opinion polls. People told the journalists themselves how much their work meant to them.

Chris Slaughter, news director of WWL-TV in New Orleans, said his station believes local television's job in a crisis like Katrina is to get information to the people, as clearly, rapidly, and accurately as possible. People need to know what has happened, which neighborhoods were most impacted, when and if to get out of the city, and where to go. He believes his station—the only television outlet in the city that stayed on the air throughout the hurricane and the subsequent flooding—did a good job at remaining focused on that function.

"We were warmly received by the public, widely lauded. They told us our coverage meant a lot to them. Their deal was, you guys stayed. You were there when we needed you."

And former WWL-TV meteorologist John M. Gumm, now working at WKRC-TV in Cincinnati, Ohio, said he received hundreds of emails and letters from viewers thanking him for convincing them to leave the city prior to Hurricane Katrina.

"People who both stayed and left told us WWL-TV was their 'lifeline' during and after the storm," Gumm said. "Tragedy usually has a way of bringing people together, and I believe that is what happened here. Our viewers know that we were going through the exact same thing they were going through; yet we were still there doing our jobs trying our best to give them information. I believe this helped them to better see us as real people who were doing all we could to give them helpful information despite the circumstances."

NBC News' Brian Williams witnessed the same sort of reaction when, six months after the hurricanes, he commented that his network planned to continue its coverage even though it was receiving criticism from some who had tired of the story.

Public reaction was instantaneous and highly supportive. Williams' blog, "The Daily Nightly," was flooded with e-mails.[1] Some were critical of the coverage, but the vast majority praised NBC's decision, complimented the coverage, raised major issues, and commented on the value of journalism in a democratic society.

"Mr. Williams, please continue to provide the people of not just the U.S., but of the world, with the continuing struggles of Katrina victims," wrote Laura Varley of Spokane, Washington. "Keep the pressure on so people cannot say they 'didn't know' in order for them to better excuse their hiding their heads in the sand."

Holding up a Mirror to Look at Ourselves

Many correspondents wrote of the continued suffering of those whose homes and lives were destroyed in the storms and stressed the need for journalism to tell these stories to the American people. "It takes courage to hold up a mirror for our nation to look at itself," wrote Dr. John Fulwiler of the Mississippi Gulf Coast. "NBC's

Brian Williams (as well as MSNBC.com) is doing that. The total destruction of the Mississippi Gulf Coast and the City of New Orleans is the truth...it is a major story that needs to be told."

By providing such coverage, said Carolyn Bain of Bethesda, Maryland, NBC (and other news organizations) "restore[s] the public's faith in unbiased media reporting. Please continue to be brave in the face of criticism....You are doing a job that few would dare to do when media channels are owned by corporations and those corporations are not well-represented by stock holders and stake holders from the New Orleans area."

And Sally Schroeder of Riverside, California, asked: "If you and NBC abandon your excellent follow-ups on Katrina, who or what will serve as the conscience for what was not done before Katrina struck as well as what is yet to be done to restore New Orleans and the other areas devastated by this storm?"

The public opinion polls confirm this public support. First of all, the polls support what everyone already knows. Television, in both 9/11 and the Gulf Coast hurricanes, was the most important source of information.[2] Americans spent hours watching the aftermath of the terrorist attacks, according to the Pew Research Center, with 81 percent saying they were constantly tuned in to news reports and 63 percent reporting being "addicted" to the coverage.[3] Networks preferred by most Americans were CNN and ABC.[4]

Likewise, with regard to the continuing stories of the devastation brought to the American Gulf Coast by hurricanes Katrina and Rita, the Public Policy Research Lab of Louisiana State University reported that nearly 78 percent of Louisianans gained their news about the impact of the hurricanes from television, followed by 8 percent who responded in favor of both newspapers and radio.[5]

Those viewers said they appreciated the journalistic work. The WestGroup Research telephone survey on September 12, 2001— one day after the attack—found that the public "overwhelmingly" approved the television coverage, and a survey by Pew Research Center from September 13-17, 2001, found "unprecedented" positive reaction (89 percent).

Studies later in the fall of 2001 confirmed the immediate impression. Scripps Howard News Service and the E. W. Scripps School of Journalism interviewed 1,131 adults in October and concluded that 91 percent found TV news coverage useful.[6]

Public Especially Appreciates State and Local Media

Five years later, Louisianans also gave local television and local newspaper good marks for covering the rebuilding process in the wake of the hurricanes. As Figure 9.1 indicates, the Public Policy Research Lab reported that 27 percent said coverage of the storm by state and local media was excellent, and another 46 percent said it was good. In the aftermath, 27 percent of respondents said local television was doing an excellent job covering the rebuilding process, while 45 percent said it was good. For local newspapers, 20 percent of respondents said the rebuilding coverage was excellent and 39 percent said it was good.

On the other hand Kirby Goidel, director of the Louisiana State University-based research group, presented results that indicate state residents were not so favorable about how the national news media were covering the impact of hurricanes Katrina and Rita. They were certain that the state's image suffered in the national media, with 63 percent saying they perceived a generally negative

Figure 9.1
Comparison of Evaluations of News Media Coverage of
Rebuilding and Hurricanes[7]

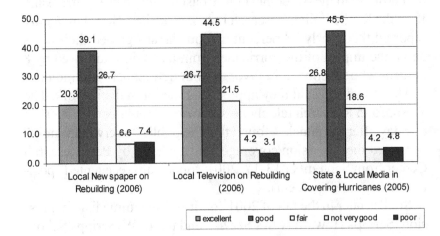

image. Curiously, however, 22 percent said the national media's coverage was excellent, and 34 percent said it was fair.

Being more specific, with 9/11 as the example, polls showed the public clearly reacted favorably to the journalistic presentation of facts. A poll by WestGroup Research showed the professionalism of the television anchors—in other words, the tone of the presentation—was mentioned by 15 percent of those surveyed as a positive in television coverage. Additionally, 36 percent mentioned accuracy of the news coverage as a positive, and about 20 percent cited the immediacy of coverage and the commitment by stations to offer continuous coverage as positives.[8]

In terms of satisfaction with 9/11 coverage, as Figure 9.2 indicates, 59 percent had no suggestions for improvement, and of those who mentioned areas that might be improved, only 15 percent at that time suggested less speculation and/or sensationalism. The WestGroup poll would indicate that television journalists had their priorities straight. They focused on what the public wanted—accurate facts, quick information, strong visuals, constant coverage, all

Figure 9.2
Suggestions for Improvement of Television Coverage of 9/11

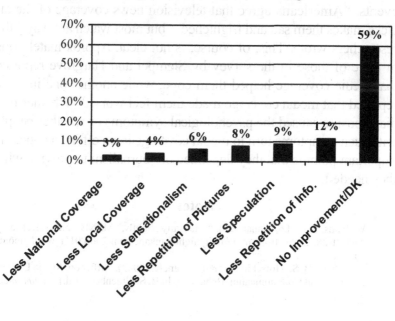

wrapped in a non-emotional tone. If television provided therapy,[9] it did so as a byproduct of its routine.

A desired result of the work of journalism, of course, is an informed public. In the case of the Gulf Coast hurricanes—with circumstances dictating that the public was forced to rely on journalism, and especially television, to find out what was happening the media seem to have been effective. The Public Policy Research Lab reported that 32 percent of its Louisiana respondents labeled themselves as very well informed and another 54 percent said they were somewhat informed.[10]

More than that, however, research confirms that the kind of intensive crisis coverage as was provided about 9/11 and hurricanes Katrina and Rita contributes to an enormous psychological impact on the viewers, listeners, and readers. The Pew Research Center, for example, reported that 71 percent reported after 9/11 they felt depressed and nearly half had difficulty concentrating. One in three had sleeping problems, and nearly seven of ten reported crying more. At least half of parents restricted young children from viewing television.[11]

The Pew researchers speculated about a possible connection between television watching and psychological reactions on the events. "Americans agree that television news coverage of the attacks makes them sad and frightened—but most watch anyway," the researchers wrote. This, of course, is not clear. Approximately one in three of those in the survey by Stempel and Hargrove reported that media coverage helped them cope, while another one in three reported that media coverage made them feel worse.[12] Whether media exposure caused the psychological symptoms or whether people with such symptoms turn to media coverage in an effort to cope, the important point is that they depended on journalism to provide what they needed.

Notes

1.   Williams, B. (2006, January 25). "The Daily Nightly," MSNBC News. Retrieved August 28, 2006, from http://dailynightly.msnbc.com/2006/01/a_word_about_ ou.html.
2.   Greenberg, B. S., Hofschire, L., & Lachlan, K. (2002). Diffusion, Media Use and Interpersonal Communication Behaviors. In B. S. Greenberg (Ed.): *Communica-*

tion and terrorism: Public and media responses to 9/11 (pp. 3-16). New Jersey: Hampton Press; WestGroup Research (2001, September 13). Americans Believe Attack on America TV Coverage Accurate; Anchors Professional, Press Release. Retrieved October 22, 2003, from http://www.westgroupresearch.com/crisiscoverage/.; Public Policy Research Lab (2006), *2006 Louisiana Survey, Reilly Center for Media & Public Affairs*. Baton Rouge, LA: Louisiana State University.

3. Pew Research Center. (2001, September 19). American Psyche Reeling from Terror Attacks, press release. Retrieved on July 2, 2007, from *http://people-press. org/reports/display.php3?ReportID=3*.

4. WestGroup Research. (2001, September 13). Americans Believe Attack on America TV Coverage Accurate; Anchors Professional, Press Release. Retrieved October 22, 2003, from http://www.westgroupresearch.com/crisiscoverage/.; Pew Research Center (2001, September 19). American Psyche Reeling from Terror Attacks.

5. Public Policy Research Lab. (2006). *2006 Louisiana Survey, Reilly Center for Media & Public Affairs*.

6. Stempel, G., & Hargrove, T. (2002). Media Sources of Information and Attitudes about Terrorism. In B. S. Greenberg (Ed.): *Communication and terrorism*, Creskill, NJ: Hampton Press.

7. Public Policy Research Lab. (2006). *2006 Louisiana Survey, Reilly Center for Media & Public Affairs*. This study was conducted between November 2005 and April 2006.

8. WestGroup Research. (2001, September 24.) Americans Believe Attack on America TV Coverage Accurate; Anchors Professional.

9. Schramm, W. (1965). Communication in Crisis. In B.S. Greenberg & E. B. Parker (Eds.): *The Kennedy assassination and the American public: Social communication in crises* (p. 1-25). Stanford, CA: Stanford University Press.

10. Public Policy Research Lab. (2006). *2006 Louisiana Survey, Reilly Center for Media & Public Affairs*.

11. Pew Research Center. (2001, September 19). American Psyche Reeling from Terror Attacks.

12. Stempel, G., & Hargrove, T. (2002). Media Sources of Information and Attitudes about Terrorism. In B. S. Greenberg (Ed.), *Communication and terrorism* (pp. 22-25). Creskill, NJ: Hampton Press.

References

Adams, C. (undated). In the wake of Katrina: Going home to a "no-man's land," [brochure]. *American University School of Communication.* Retrieved September 8, 2007, from http://147.9.1.138/main.cfm?pageid=1365.

Adams, L. (2005, September 2). *Why some media coverage of Katrina is patently offensive.* Retrieved February 23, 2006, from http://www.authorsden.com/visit/viewarticle.asp?AuthorID=2517&id=19339.

Bergeron, K. (2006). *Leadership roundtable video presentation.* Quote extracted from the 2006 Associated Press Managing Editors Conference, New Orleans, LA.

Brasch, W. M. (2006). *The federal response to Hurricane Katrina.* Charleston, SC: BookSurge.

Breed, W. (1980). *Dissertations on sociology: The newspaperman, news, and society.* New York: Amo Press.

Brinkley, D. (2006). *The great deluge: Hurricane Katrina, New Orleans, and the Mississippi Gulf Coast.* New York: Harper Perennial.

Cancel, A., Cameron, G., Sallot, L., & Mitrook, M. (1997). It depends: Contingency theory of accommodation in public relations. *Journal of Public Relations Research, 9*(1), 31-63.

Chin, O. S. (2006, January 7). In 2005, journalists became the news. *The Straits Times, Singapore.* Retrieved on September 8, 2006, from http://www.asiamedia.ucla.edu/article-world.asp?parentid=36786.

Cooper, A. (2005, September 1). 360 Special Edition, CNN. Retrieved May 1, 2009, from http://transcripts.cnn.com/TRANSCRIPTS/0509/01/acd.01.html.

Cooper, A. (2006, June 1). *Larry King Show,* CNN Transcripts. Retrieved April 13, 2008, from www.cnn.com/transcripts

Cooper, C., & Block, R. (2006). *Disaster: Hurricane Katrina and the failure of homeland security.* New York: Times Books, Henry Holt.

Cough, P.J. (2007, May 25). The case of the disappearing TV viewers, Reuters. Retrieved April 22, 2009, from http://reuters.com/article/entertainmentNews/idUSN2523545420070525.

Coy, P., Foust, D., Woellert, L., & Paleri, C. (2005, September 12). Katrina's wake. *Business Week, 32.*

Crichton, Michael. (1993), Mediasaurus: The Decline of Conventional Media, speech to the National Press Club, Washington, D.C., April 7, 1993. Retrieved April 22, 2009, from http://www.crichton-official.com/speech-mediasaurus.html

Dart Center for Journalism and Trauma (2005, September). A sense of outrage: Covering Katrina's aftermath. Retrieved September 8, 2007, from http://www.dartcenter.org/articles/personal_stories/hewitt_gavin.html

Deggans, E. (2005, September 8). Journalists' Outrage Visible in Coverage. *St. Petersburg Times,* 6A.

de Zengotita, T. (2002, April). The numbing of the American mind: Culture as anesthetic. *Harper's Magazine.* Retrieved September 8, 2007, from http://www.csubak. edu/~mault/Numbing% 20of%20american%20mind.htm.

Dixon, Travis (undated). Understanding News Coverage of Hurricane Katrina: The Impact of News Frames and Stereotypical News Coverage on Viewer's Conceptions of Race and Victimization. Retrieved May 4, 2009, from http://katrinaresearchhub. ssrc.org/understanding-news-coverage-of-hurricane-katrina-the-impact-of-news-frames-and-stereotypical-news-coverage-on-viewers-conceptions-of-race-and-victimization/project_view.

Dyson, M. E. (2006). *Come hell or high water: Hurricane Katrina and the color of disaster.* New York: Basic Civitas Books.

Entman, R. (1993). Framing: Toward Clarification of a Fractured Paradigm, *Journal of Communication, 43 (4),* Pg. 51-59.

Entman, R., & Rojecki, A. (1993). Freezing out the Public: Elite and Media Framing of the U.S. Anti-nuclear Movement, *Political Communication, 10 (2),* Pg. 155-173.

Gabler, N. (2005, October 9). Good night, and the good fight. *New York Times,* 12.

Gans, H. J. (1979). The messages behind the news. *Columbia Journalism Review,* January-February, 40-45.

Gans, H. J. (1979). *Deciding what's news: A study of CBS Evening News, NBC Nightly News, Newsweek and Time.* New York: Pantheon Books.

Goldberg, J. (2007, September 5). Storm of malpractice: Katrina was a media disaster. *National Review Online.* Retrieved September 8, 2007, from http://article.nationalreview.com/print/?q=NmEyNj MzMWQ1OTI3ZjhiMmE5YWNkZDc2Mm-M2NDQ1NTg=.

Goss, A., Harwood, B., McElroy, P., & Baker, C. (2006, May 12). Profile: Katrina photojournalist John McCusker. *Voices of New Orleans.* Retrieved on September 12, 2006, from http://www.prx.org/pieces/12547.

Grace, S. (2007, January 23). Brown's troubling remarks deserve airing. *Times-Picayune: Metro editorial,* 5.

Greenberg, B. S., Hofshire, L., & Lachlan, K. (2002). Diffusion, media use and interpersonal communication behaviors. In B.S. Greenberg (Ed.), *Communication and terrorism: Public and media responses to 9/11* (pp. 3 – 16). Cresskill, NJ: Hampton Press.

Grunig, J. (1997). Public relations management in government and business. In J. Garnett & A. Kouzmin (Eds.), *Handbook of administrative communication.* New York: Marcel Dekker, Inc.

Grunig, J., & Grunig, L. (1990). Models of public relations: A review and reconceptualization. Paper presented to the Association for Education in Journalism and Mass Communication, Minneapolis, MN.

Grunig, J., & Hunt, T. (1984). *Managing public relations.* New York: Holt, Rinehart and Winston.

Hamilton, J.M., Burnett, J.F., Wells, K. & Thompson, I. (2006, October 28). Journalists report: Katrina and her aftermath. Symposium conducted at the Louisiana Book Festival Panel Discussion, Baton Rouge, LA.

Hannity, S. (2005, September 6). Special edition: Hurricane Katrina. *The O'Reilly Factor.* Fox News Transcripts.

Hannity, S., & Colmes, A. (2007, September 7). Should New Orleans be rebuilt? *Fox News Transcripts.*

Harden, B., & Moreno, S. (2005, September 23). Thousands fleeing Rita jam roads from coast. *The Washington Post,* A1.

Hartman, C., & Squires, G. D. (Eds.) (2006). There is no such thing as a natural disaster.

New York: Routledge (Taylor & Francis Group).

Haynes, M. (2006, December 22). A tragedy's emotional impact can engulf journalists. *Pittsburgh Post-Gazette*, E1.

Horne, J. (2006). *Breach of faith: Hurricane Katrina and the near death of a great American City.* New York: Random House.

Hume, B. (2005, September 5). Who's to blame and who isn't to blame. *Special Report with Brit Hume.* Retrieved on April 13, 2008, http://www.foxnews.com/story/0,2933,168582,00.html.

Hume, B. (2005, September 11). *Fox News Sunday,* Fox News Transcripts. Retrieved on April 13, 2008, from http://mediamatters.org/items/200509120001.

Katrina Research Hub, Social Science Research Council. Available at http://katrinaresearchhub.ssrc.org/.

Knickerbocker, B., & Jonsson, P. (2005, September 8). New Orleans toxic tide. *Christian Science Monitor,* A1.

Kushner, A. B. (2005, September 12). After the flood. *The New Republic,* 42.

Kurtz, H. (2007). *Reality show: Inside the last great television news war.* New York: Free Press.

Lang, D. (2007, December 13). Katrina Photographer McCusker Fined, Placed on Probation. *PDNonline.* Retrieved on March 8, 2008, from http://www.pdnonline.com/pdn/newswire/article_display.jsp?vnu_content_id=1003685185

Lichtenberg, J. (2000). In defence of objectivity revisited. In J. Curran & M. Gurevitch (Eds.), *Mass media and society* (pp. 238-254). London: Arnold.

Lipton, E. (2006, February 10). White House knew of levee's failure on night of the storm. *New York Times,* A1.

Lyon, L., & Cameron, G. (2004). A relational approach examining the interplay of prior reputation and immediate response to a crisis. *Journal of Public Relations Research, 16*(3): 213-241.

Marra, F. (1998). Crisis communication plans poor predictors of excellent public relations. *Public Relations Review, 24*(4), 461-474.

McFadden, K. (2005, December 30). Emotional surrender; The sci-fi invasion fizzled while TV on-demand soared, but passion was everywhere in 2005. *Seattle Times,* I1.

Media Matters for America (2005, September 11). Knight Ridder, Palm Beach Daily News, Fox News' Hume repeated misleading Red Cross story, shifted blame to Blanco. *Palm Beach Daily News.* Retrieved on April 13, 2008, from http://mediamatters.org/items/200509120001.

Mindich, D. T. (1998). *Just the facts: How objectivity came to define American journalism.* New York: New York University Press.

Moritz, Marquer (undated). Covering Katrina: The Multiple Dilemmas of Local Journalists. Retrieved May 4, 2009, from http://katrinaresearchhub.ssrc.org/covering-katrina-the-multiple-dilemmas-of-local-journalists/project_view.

Mulrine, A., Marek, A. C., & Brush, S. (2005, September 12). To the rescue. *U.S. News & World Report, 39* (9), 20-26.

Nagourney, A., & Kornblut, A. E. (2005, September 5). White House enacts a plan to ease political damage. *The New York Times,* 14.

National Research Council Committee on Disasters and the Mass Media. (1980). *Disasters and the mass media: Proceedings of the committee on disasters and the mass media workshop.* Washington, DC: National Academy of Sciences.

NOAA Technical Memorandum NWS TPC-1. (1997, February). The deadliest hurricanes in the United States, 1900-1996. Retrieved September 8, 2007, from http://www.nhc.noaa.gov/pastdead.html.

O'Reilly, B. (2005, September 6). *The O'Reilly Factor*, Fox News Transcripts. Retrieved on April 13, 2008, from http://mediamatters.org/items/200509120001.

Overholser, G. (2006). On behalf of journalism: A manifesto for change. Retrieved on September 1, 2007, from http://www.annenbergpublicpolicycenter.org/Overholser/20061011_JournStudy.pdf .

Overholser, G. (2008). Updating "A Manifesto for Change." Manship School of Mass Communication's Reilly Center for Media & Public Affairs. Retrieved April 30, 2009, from http://www.genevaoverholser.com/?q=node/14.

Parter, D. (2006). *Leadership roundtable video presentation.* Quote extracted from the 2006 Associated Press Managing Editors Conference, New Orleans, LA.

Pew Research Center for the People & the Press (2001, September 19). American psyche reeling from terror attacks [Press Release]. Retrieved August 22, 2007, from http://people-press.org/reports/display.php3?ReportID=3.

Public Policy Research Lab (2006), *2006 Louisiana survey, Reilly Center for Media & Public Affairs.* Baton Rouge, LA: Louisiana State University.

Quarantelli, E.L. (1991, January). Lessons from research: Findings of mass communication system behavior in the pre, trans, and postimpact periods of disasters; Preliminary Paper #160. Paper presented at the meeting of Crisis in the Media, Emergency Planning College, York, England.

Quarantelli, E. L., & Dynes, R. R. (1970). Property norms and looting: Their patterns in community crises. *Phyon, 31*, 168-182.

Reese, S. D., & Danielian, L. J. (1989). Intermedia influence and the drug issue: Converging on cocaine. In P. J. Shoemaker (Ed.), *Communication campaigns about drugs, media and the public,* (pp. 29-45). Hillsdale, NJ: Lawrence Erlbaum Associates.

Roig-Franzia, M., & Hsu, S. (2005, September 5). Many evacuated, but thousands still waiting, corrections. *The Washington Post,* A2.

Rosen, J. (2004). Our code is falling to pieces: Doug McGill on the fading mystique of an objective press. *PressThink.* Retrieved on April 13, 2008, from http://journalism.nyu.edu/pubzone/ weblogs/pressthink/2004/10/29/mcgill_essay.html.

Schramm, W. (1965). Communication in crisis. In B.S. Greenberg & E. B. Parker (Eds.), *The Kennedy assassination and the American public: Social communication in crises.* Stanford, CA: Stanford University Press.

Seelye, K., Carter, B., & Elliot, S. (2005, September 12). To publish, not perish: New Orleans news media soldier on, ad-free. *The New York Times,* p. C1.

Shane, S., & Lipton, E. (2005, September 2). Government saw flood risk by levee failure. *New York Times,* A1.

Staff Reports (2006, August 9). N.O. Man arrested after chase; He asked cops to shoot him, police say. *The Times-Picayune,* B1.

Stempel, G., & Hargrove, T. (2002). Media sources of information and attitudes about terrorism. In B. S. Greenberg (Ed.), *Communication and Terrorism* (pp. 3-16). Cresskill, NJ: Hampton Press.

Stephens, K., Malone, P., & Bailey, C. (2005). Communicating with stakeholders during a crisis: Evaluating message strategies. *Journal of Business Communication, 42*(4): 390-419.

Stephanopoulos, G. (2005, September 11). Relief effort New Orleans Update. *ABC News* Transcript.

Strupp, J. (2006, August 9). Times-Pic Editor comments on arrest of photographer, *Editor & Publisher.* Retrieved September 13, 2006, from http://editorandpublisher.com.

Strupp, J. (2006, August 10). More than $9,000 raised for arrested "Times-Picayune" photographer. *PDNonline.* Retrieved March 8, 2008, from http://www.pdnonline.

com/pdn/newswire/article_display.jsp?vnu_content_id=1002985303.

Sylvester, J.(2008). *The Media and Hurricanes Katrina and Rita.* New York: Palgrave Macmillan.

Thevenot, B., & Russell, G. (2005, September 26). Rumors of death greatly exaggerated, *The Times-Picayune.* Retrieved April 13, 2008, from http://www.nola.com/newslogs/tporleans/index.ssf?/mtlogs/nola_tporleans/archives/2005_09_26.html.

Thomas, M. (2005, September 12). The lost city. *Newsweek,* 42.

Times-Picayune, New Orleans, Staff Reports. (1006, August 9). N.O. man arrested after chase. He asked cops to shoot him, police say. *The Times-Picayune,* p. B1. Retrieved September 13, 2006, from http://www.nola.com.

Toosi, N. (2007, January 20). Former FEMA head Brown says party politics played role in Katrina response. *Associated Press Worldstream.*

United States Congress (2006, February 19). *A failure of initiative: Final report of the select bipartisan committee to investigate the preparation for and response to Hurricane Katrina.* Washington, D.C.: Government Printing Office.

Valkenburg, P. M., Semetko, H. A., &. DeVreese, C. H. (1999). The effects of news frames on readers' thoughts and recall. *Communication Research, 26*(5), 550-569.

Varney, J. (2006). Leadership roundtable video presentation. Quote extracted from the 2006 Associated Press Managing Editors Conference, New Orleans, LA.

Wenger, D. E. (1985). Mass media and disasters; Preliminary paper No. 98. Newark, DE: Disaster Research Center, University of Delaware.

Warner, C. (2007, January 17). Small talk; Living with the family in a FEMA trailer isn't a disaster, but the close quarters and lack of privacy won't be missed, *The Times-Picayune,* p. 1.

WestGroup Research. (2001, September 13). Americans believe attack on American TV coverage accurate; anchors professional [Press Release]. Retrieved October 22, 2003, from http://www.westgroupresearch.com/crisiscoverage/.

The White House. (2006, February). *The Federal Response to Hurricane Katrina: Lessons Learned.* Retrieved on April 13, 2008, from http://www.whitehouse.gov/reports/katrina-lessons-learned.pdf.

Williams, B. (2006, January 25). The Daily nightly, MSNBC News. Retrieved August 28, 2006, from http://dailynightly.msnbc.com/2006/01/a_word_about_ou.html.

Williams, B. (2006, August 28). Brian Williams: We were witnesses. *NBC News.* Retrieved September 9, 2007, from http://rss.msnbc.msn.com/id/14518359/.

Williams, B. (2005, Sept. 15). The weatherman nobody heard. *NBC News.* Retrieved September 9, 2007, from http://www.msnbc.msn.com/id/9358447/.

Williams, B. (2006, August 28). Katrina, the long road back; One-year anniversary. *NBC Nightly News.*

List of Contributors

Jinx C. Broussard is associate professor, Manship School of Mass Communication, Louisiana State University.

Jane Dailey is assistant professor, Communication and Media Studies, Marietta College.

Roxanne K. Dill is instructor in the Manship School of Mass Communication, Louisiana State University.

Ralph Izard is Sig Mickelson/CBS professor in the Manship School of Mass Communication, Louisiana State University, and professor emeritus, E.W. Scripps School of Journalism, Ohio University.

Lisa K. Lundy is associate professor, Manship School of Mass Communication, Louisiana State University.

Robert Mann is Manship Chair, Manship School of Mass Communication, Louisiana State University. At the time of Hurricane Katrina, he was communications director for Louisiana Governor Kathleen Babineaux Blanco.

Kirsten Mogensen is associate professor, Department of Communication, Business and Information Technologies, Roskilde University, Denmark.

Jay Perkins is associate professor, Manship School of Mass Communication, Louisiana State University.

Shearon Roberts is a freelance journalist in Miami, Florida.

Guido H. Stempel III is distinguished professor emeritus, E.W. Scripps School of Journalism, Ohio University.

List of Interviews*

Jim Amoss, Executive Editor, the *Times-Picayune*, New Orleans, September 21, 2006

Kathy Anderson, Photographer, the *Times-Picayune*, New Orleans, September 21, 2006

John Balance, Photo Department Manager, *The Advocate*, Baton Rouge, La., September 1, 2006

Kim Chatelain, Suburban Editor, the *Times-Picayune*, New Orleans, April 11, 2006

Art Daglish, *Cox Newspapers*, Washington, April 21, 2006

John DeSantis, Reporter, the *New York Times*, September 13, 2006

Frank Donze, City Government Reporter, the *Times-Picayune*, New Orleans, September 21, 2006

John M. Gumm, Meterologist, WKRC-TV, Cincinnati, September 7, 2007

Kendall Hebert, Director of Public Relations, Louisiana Capital Area Chapter of the American Red Cross, Fall 2005

Mike Hoss, Anchor, WWL-TV, New Orleans, November 16, 2006

Ted Jackson, Senior Photographer, the *Times-Picayune*, New Orleans, April 11, 2006

Note: Many of these journalists have changed jobs since they were interviewed. They are listed here in the position they held at the time of the interview.

Bob Johanessen, Communications Director, Louisiana Department of Health and Hospitals, Fall 2005.

Blake Kaplan, Assistant City Editor, Biloxi (Miss.) *Sun Herald*, September 8, 2006

John Kieswetter, TV/Media Writer, *Cincinnati Enquirer*, September 7, 2006

Peter Kovacs, Managing Editor, the *Times-Picayune*, New Orleans, August 1, 2006

Anita Lee, Reporter, Biloxi *Sun Herald*, September 5, 2006

Trymaine Lee, Police Reporter, the *Times-Picayune*, New Orleans, September 21, 2006

Linda Lightfoot, Executive Editor, *The Advocate*, Baton Rouge, La., August 8, 2006

Larry Lovell, Account Supervisor, Peter Mayer Public Relations, New Orleans, Fall 2005

Spud McConnell, Radio Talk Show Host, WWL-870 AM, New Orleans, November 16, 2006.

David Meeks, City Editor, the *Times-Picayune*, New Orleans, April 11, 2006; September 21, 2006

Tim Morris, State/Political Editor, the *Times-Picayune*, New Orleans, April 11, 2006

Robert Pierre, Reporter, *Washington Post*, August 13, 2006

Gary Raskett, Circulation Manager, Biloxi (Miss.) *Sun Herald*, March 2, 2007

Carl Redman, Managing Editor, *The Advocate*, Baton Rouge, La., August 22, 2006

John Roberts, Senior National Correspondent, CNN, April 21, 2006

Gordon Russell, City Government Reporter, the *Times-Picayune*, New Orleans, September 21, 2006

Mark Schleifstein, Environmental Reporter, the *Times-Picayune*, New Orleans, September 21, 2006

Dan Shea, Managing Editor, News, the *Times-Picayune*, New Orleans, April 11, 2006

Debra Skipper, Assistant Managing Editor for News and Business, the *Clarion-Ledger*, Jackson, Miss., September 15, 2006

Chris Slaughter, News Director, WWL-TV, New Orleans, September 6, 2007

Chris Spencer, Senior Communications Manager, JPMorgan Chase, Baton Rouge, Louisiana, Fall 2005.

Manuel Torres, Assistant City Editor, the *Times-Picayune*, New Orleans, April 11, 2006

Terri Troncale, Op-Ed Page Editor, the *Times-Picayune*, New Orleans, September 21, 2006

Brian Williams, Anchor and Managing Editor, *NBC Nightly News*, November 16, 2006

Index

Printed in the United States
by Baker & Taylor Publisher Services

Printed in the United States
by Baker & Taylor Publisher Services